HowEx Guide to Recycling

101+ Tips to Learn How to Recycle, Eliminate Disposables, Reduce Waste & Pollution, Conserve Resources, Save Energy, and Protect the Environment

HowExpert with Jen Thilman

For more tips related to this topic, visit HowExpert.com/recycling.

Recommended Resources

- HowExpert.com – How To Guides on All Topics from A to Z by Everyday Experts.
- HowExpert.com/free – Free HowExpert Email Newsletter.
- HowExpert.com/books – HowExpert Books
- HowExpert.com/courses – HowExpert Courses
- HowExpert.com/clothing – HowExpert Clothing
- HowExpert.com/membership – HowExpert Membership Site
- HowExpert.com/affiliates – HowExpert Affiliate Program
- HowExpert.com/jobs – HowExpert Jobs
- HowExpert.com/writers – Write About Your #1 Passion/Knowledge/Expertise & Become a HowExpert Author.
- HowExpert.com/resources – Additional HowExpert Recommended Resources
- YouTube.com/HowExpert – Subscribe to HowExpert YouTube.
- Instagram.com/HowExpert – Follow HowExpert on Instagram.
- Facebook.com/HowExpert – Follow HowExpert on Facebook.
- TikTok.com/@HowExpert – Follow HowExpert on TikTok.

Publisher's Foreword

Dear HowExpert Reader,

HowExpert publishes quick 'how to' guides on all topics from A to Z by everyday experts.

At HowExpert, our mission is to discover, empower, and maximize everyday people's talents to ultimately make a positive impact in the world for all topics from A to Z...one everyday expert at a time!

All of our HowExpert guides are written by everyday people just like you and me, who have a passion, knowledge, and expertise for a specific topic.

We take great pride in selecting everyday experts who have a passion, real-life experience in a topic, and excellent writing skills to teach you about the topic you are also passionate about and eager to learn.

We hope you get a lot of value from our HowExpert guides, and it can make a positive impact on your life in some way. All of our readers, including you, help us continue living our mission of positively impacting the world for all spheres of influences from A to Z.

If you enjoyed one of our HowExpert guides, then please take a moment to send us your feedback from wherever you got this book.

Thank you, and we wish you all the best in all aspects of life.

Sincerely,

Byungjoon "BJ" Min
Founder & Publisher of HowExpert
HowExpert.com

PS...If you are also interested in becoming a HowExpert author, then please visit our website at HowExpert.com/writers. Thank you & again, all the best!

Table of Contents

Introduction

I'm excited you picked up this book. You see the need to do what you can for our planet. But, as I unfold the story of how we got here, I think you'll agree that the real blame falls on ignorance and greed. They have led to the disasters we see today, like excessive fires, flooding, and drought.

- Silently, disposable habits and overconsumption crept into our lives before we realized the damage they were causing.

How long can the planet handle our disposable lifestyle?

Look at the news or hear about environmental issues, and climate anxiety is bound to hit you. Many of us experience it. I did researching this book. We're more anxious about a problem when we feel we can't do anything about it.

We control how we eat, what we buy, and how much we waste.

I've heard people say, "It's the big corporations; they're the ones who need to change." While this is true, I don't see that as a reason not to reduce our waste and make smarter, healthier buying decisions.

Pointing blame won't get us out of this mess.

It's time we move forward and build a better world.

Corporations will change when we use our wallets and voices to make them change.

This book is intended to be actionable. It has tips to help reverse and stop the damage we've done. My hope is it will become a tool you keep on hand and use regularly.

Bookmark the pages related to what you are working on at any one time.

Keep reading tips to build habits that reduce your environmental impact.

You may not be able to do everything in this book right now, like buy an electric car, but you can change how you consume resources like energy, food, and water.

Look for actions you can do today, like turning off lights and faucets or learning eco-friendly ways to shop.

Work up to more changes you want to make in the future.

You're the only one who can change your habits.

Knowledge is power. This book will help you gain knowledge so you feel empowered to build a healthier planet. I'm working to build these habits. I hope you'll join me.

We're all in this together, so let's fix it together.

Chapter 1: Plastics - Our Biggest Environmental Threat

One of the biggest contributors to the climate crisis is single-use plastics. So if you do nothing else, please minimize your use of these.

Though looking at how we reduce, reuse, and recycle all disposables is important, plastics must be the priority for several reasons.

- The sheer volume of plastics that have been created.
  - Since the 1950s, over nine billion tons of plastics have been produced, and though we may have thought they were being recycled, they weren't.
- Plastics are difficult to recycle. That may sound strange since we've put plastics in our recycling bins for years, but that doesn't mean they've been recycled.
  - Over 90% of single-use plastics have ended up in landfills and oceans or were incinerated.
  - In 2021, Americans disposed of over 300 pounds of plastics per person.
- Plastics are made from fossil fuels and significantly increase damaging emissions.
  - This is especially true with single-use plastics, which continue to cause emissions and environmental damage when discarded.

Tip 1 - Putting plastics in a recycling bin doesn't mean they get recycled.

Since household recycling started in the 1970s, only 9% of plastics have been recycled into another container or item. These were either upcycled into a non-disposable item or downcycled into something that ended up as trash. The biggest reason for this is that most plastics aren't made to be recycled in the first place.

- Most single-use plastics sold in the US don't meet the quality standards required to be recycled.
- A plastic soda bottle isn't usually turned into another bottle.

- When plastic is recycled, it loses strength and can't be used to create the same type of plastic.
- If they do get recycled, plastic bottles are often used to make lower-grade plastics that aren't strong enough to be recycled again.
  - This is called downcycling.
- Some plastics are upcycled and turned into non-disposable items of greater value.
  - Recycled polyester in clothing is usually from upcycled plastic bottles.

Other reasons that plastics don't get recycled are...

- Non-recyclable plastics, like bags or other soft plastics, and food waste put in bins cause contamination, so materials that can be recycled end up as trash.
- Not enough manufacturers are using recycled packaging, so the demand for post-consumer plastics is low.

The softer the plastic, the less likely it can go in your bin.

Most services don't recycle the materials we give them; they only sort and sell them. They're what's called a Materials Recovery Facility (MRF). They stay in business by selling recyclables on the commodities market.

- Even if they take 3 - 7 plastics, the market to sell these is minimal.
  - There aren't many buyers for plastics other than 1's and 2's.
    - In fact, 3's and 4's are soft plastics like bags, 6's are polystyrene foam (StyrofoamTM), and 7's are *other* plastics that can't be recycled.

It's not easy to figure out what you can put in your bin, even if you look on the website of your recycling service or MRF.

I looked at the plastics recycling page for a major waste management company. They say they accept 1 - 7 plastics but went on to say they only take bottles, milk jugs, and hard plastics like detergent and shampoo bottles without specifying numbers. So, they only take 1's and 2's, but they don't make that clear. *How confusing!*

The Life of Plastics - Where Does It All Go?

The recycling triangle was introduced in 1988 by the plastics industry. Originally, the stamp identified the 1's and 2's that had the *potential* to be recycled, even though they may not.

MRFs pushed back against the triangle since plastics increased their costs, and there were no buyers for this material. But producers kept making more and different types of plastics, and we kept trying to recycle them. They have a triangle, after all.

We would have been better off if the recycling triangle on plastics meant something.

The numbers on plastics were developed to give us the *impression* they can be recycled, even though the plastics industry knew this wasn't true. It was better for their business if we believed it, and we wanted to believe it because plastics are so convenient.

If something makes life easier, and we don't need to spend a lot of money on it, it makes sense to do it.

If the recycling triangle had been regulated so it was only on plastics that could be recycled, many of us would have avoided the ones that couldn't, and a lot less would be made today.

Tip 2 - Make a list of single-use plastics to eliminate or reduce using them.

Single-use plastics are a silent killer we barely notice. A way to reduce them is to figure out where they come into our lives. Look at the disposable plastics you buy and how to get rid of them. Most will be in your kitchen and bathroom.

- Write them down or save paper and do it on your phone or computer. List everything you regularly buy that comes packaged in plastic. It's likely to be a lot.
  - List the groceries, household items, electronics, and anything that comes with some form of plastic packaging - either wrapped in it or shipped with it. Then highlight the things you can change how you buy or use.

Hint: If you use plastic shopping bags, list and highlight these. This is one of the easiest plastics to eliminate and should be one of the first you tackle. **See Tip 7.**

- Keep the list going, adding to it as you think of things.
- Make it a checklist. It feels good to check things off a list.

Throughout this book, you will find ways to reduce, reuse, and properly recycle plastics. Knowing where they come from will make it easier.

Tip 3 - Give up or reduce single-use plastics to save money and the planet.

It's scary that over half of the plastics currently being produced in the world are for single use. But, unfortunately, the convenience of these has so invaded our lives that most people barely notice how much plastic waste they generate.

You may not be able to completely control the type of container your purchases come in, but you can eliminate or reduce buying disposable plastics when there are alternatives.

- Switch to bar soap or reusable soap dispensers.
  - Bar soap is more environmentally friendly.
    - The impact of liquid soap is ten times worse when you consider the plastic packaging.
  - If you prefer to use liquid soap, get a refillable dispenser at a local thrift store or yard sale, and buy economy size to refill it.

- o Did you know you can buy shampoo and conditioner in a bar? Try it. You may like these better.
- Buy big bottles of detergents and cleaners or eco-friendly brands that offer concentrates and sustainable packaging. Look online for these alternatives.
  - o If your favorite brand doesn't offer eco-friendly packaging, contact the company to tell them you want this.
- If you like soft drinks, consider buying them in aluminum cans or glass bottles. These have a much better chance of being recycled.
  - o If you drink lots of soda, get a home soda maker and eliminate single-use drink containers completely.

Tip 4 - Always wash plastics and make sure to recycle them correctly.

Though the goal is to reduce buying single-use plastics, when recycling them, make sure to do it right.

- Only put what you know is accepted in your bin, and always ensure it's clean and dry.
  - o This prevents contamination of other materials that are more easily recycled.
- Next to plastics, food is one of the biggest contaminators of our recycling.
  - o Never put anything in your bin that has food or grease on it.

While thoroughly washing your recycling can be easy, figuring out what plastics go in your bin is challenging.

So Many Types of Plastic - Tips on Recycling them Correctly

When we started recycling plastics, I believed a bottle I washed and put in the bin would come back around as another bottle. That all plastics would be in a never-ending cycle of reuse, and virgin plastics would no longer be made. That was the goal of recycling, wasn't it? To eventually stop making new. *But, boy, was I naive!*

For years we believed that taking care of plastics meant looking at the bottom of a container to see if it had a triangle with a number and, if it did, washing it and tossing it in a recycle bin.

Unfortunately, plastics recycling is more complicated than that. The triangle now appears on most plastics, even if they aren't recyclable. Unfortunately, this has led us to put a lot of them in the bin that don't belong.

A numbers game - what the recycling stamp on plastics really means.

When plastics recycling started, only two types could go in bins: PET (1's), usually beverage bottles and some food containers, and HDPE (2's), heavier plastics like detergent and shampoo bottles. These two types are still the only plastics that are likely to get recycled.

Tip 5 - Just because it is plastic doesn't mean it goes in your recycling.

With the triangle showing up everywhere, we've developed a practice of hoping all plastics can be recycled. This is called wish-cycling, where we put plastics in the bin whether we know if they belong there or not. This causes problems for MRFs and increases costs to consumers.

Only certain disposable plastics belong in recycling bins.

The practice of throwing everything plastic into recycling has become so bad that New York City's recycling service receives around 1,200 bowling balls a year. *Bowling balls! Really?* Just because something's made of plastic doesn't mean it's recyclable.

Here are some plastic items commonly found in recycling bins that don't belong:

- **Soft plastics** - there are so many in our lives that it's hard to list them all.
  - Shopping, produce, and trash bags.
  - Plastic wrappers and straws.
  - Freshness bags from dry goods, like cereal or crackers.
  - Snack packaging, like chip bags and candy wrappers.
  - Juice and baby food pouches.
  - Ketchup and sauce packets.
  - Mesh produce bags your potatoes or avocados come in.
  - Plastic rings from six-packs.
- **Plastic utensils** - no forks, knives, spoons, etc.
  - If compostable, they go in the compost, not recycling.
- **Small plastics** – carry-out dressing and dip cups and their lids are good examples. They go in the trash even if they have a number and recycling triangle.
  - Small containers can't be detected by sensors in sorting.
  - Cotton swabs with a plastic core are not recyclable. They're mixed materials and too small for sorting sensors.
- **Electronics** - contain plastics not made to be recycled and are mixed with other materials. These should only be disposed of at e-waste collection sites or events.
  - They can't go in the trash. They contain chemicals and other materials we shouldn't put in the ground.
  - We need the precious metals in them to build more electronics.
- **Mixed plastics** - these disposables contain other materials besides plastic and won't get recycled in your bin.
  - Carry out coffee cups.
  - Plastic-lined delivery envelopes like bubble mailers.
  - Disposable lighters.
  - Toothbrushes - there are mail-in programs that recycle these.
  - Diapers - clean or dirty, these don't belong.
  - Masks and face shields.

- **Reusable containers** - when these wear out or break, they go in the trash.
  - Only plastics made to be disposable can potentially be recycled.
  - These are treated plastics that can't be made into another container.

Never put your recycling in a plastic bag. It needs to go in your bin loose or in a paper bag. Most MRFs will not tear open plastic bags since this can be dangerous for workers.

- There are rare exceptions, such as in New York City, where there is limited space for bins, so people put their recycling in clear plastic bags. *I bet those bowling balls bust out of the bottom!*

Tip 6 - The problem with small plastics; putting caps on bottles and reusing travel containers.

Size matters when it comes to the plastics you put in your bin. Small items like plastic caps and travel-size containers aren't big enough to be picked up by the sensors in sorting, even if they have a triangle and a number.

- Hang on to hotel shampoo containers and refill them at home for future travels.
- Buy travel sizes of your favorite brand and keep refilling them.
  - Years ago, I found my shampoo and conditioner brand in travel size. I bought one of each and still have them to refill for traveling.
- Sensors that sort plastics recognize big items like bottles, not caps.
  - Always put plastic caps back on bottles to increase the chance they'll get recycled.
  - Loose bottle caps end up in the trash.

These things can often change without us knowing. I went a few years doing it wrong before I learned that my MRF wanted caps on bottles. You miss one notice, or your MRF doesn't communicate a change, and you end up doing it wrong. It's frustrating.

What to do about shopping bags and other soft plastics.

One of the biggest culprits of recycling contamination is soft plastics, the worst of these being shopping bags. They're difficult to recycle, so MRFs don't usually take them. Because they're a big problem, legislation to stop the use of plastic shopping bags is increasing.

- Around **5 trillion** plastic bags are made annually.
- Eight states in the US had laws to control plastic bag use at the beginning of 2021, as well as over 100 countries around the globe.

We can't wait for legislation to take care of this problem. Eliminating our use of shopping bags is the quickest way to get rid of these planet killers.

Tip 7 – Eliminate plastic shopping bags; these soft plastics can't go in recycling bins.

Reusable shopping bags are cheap and can be purchased at stores or online. When these first came out, I would hear people say they often left them at home. Here are some ways to remember your bags and avoid plastics at checkout.

1. Train yourself to remember your bags; turn around and go back home for them. You're not as likely to forget them in the future.
2. Make it a habit to put them back in your car after every use.
3. Add *reusable bags* to your shopping list to remind you to grab them.
   a. Most of us look at our shopping list before we walk out the door.

- Buy compact reusable shopping bags. I discovered Baggu Bags and Chico Bags, which I love because they're small and fold up into my purse or pocket. I always have my bags with me.
  - They're durable and inexpensive. I've had mine for almost twenty years.
- If you land at the checkout without your reusable bags, ask for paper. It's less damaging to the planet, and with a healthy planet, we will always be able to grow more trees.
  - Paper bags can be easily recycled in your bin, but plastic bags can't.

While it's critical that we eliminate using plastic bags, it's equally important that we take time to recycle what we have correctly.

- The only chance plastic bags have at getting recycled is to take them to a retail store with a collection bin. There are several big chains that have these.
  - Some store collection bins accept other soft plastics, like wrappers from bathroom tissue.
  - In many areas, it is required by law that retailers take back plastic shopping bags.
- Visit Earth911.com to find drop-off locations for plastic bags in your area.
- Go to bagandfilmrecycling.org to learn more and find where to take them.

Tip 8 - Don't use plastic produce and meat bags; reusable ones are inexpensive.

Did you ever notice how thin and flimsy produce and meat bags are? These thin plastics aren't recyclable. They're a weak material that can't be used again to make more plastics.

- Instead of using plastic bags for produce or meat, buy reusable ones. These mesh bags are easy to wash, come in lots of sizes, and are inexpensive.
- Skip bagging it – no one says you must put your produce in a bag.

- Combine produce into one bag; they don't need to be bagged separately.

Tip 9 - Where to find soft plastics in your life and how to eliminate them.

Shopping bags aren't the only soft plastics we use. There are many ways this hard-to-recycle material shows up in our lives.

We use soft plastics to store food, and it comes wrapped in them.

It shows up as carry-out food bags, trash bags, and wrappers from household items like paper towels and tissue.

Most of these can and need to be eliminated.

Some of these can be recycled with plastic shopping bags in store collection bins. But soft plastics are still difficult to recycle, so it may not happen even if we put them in the proper collection bin. So we need to reduce using them whenever we can.

It's hard to change the way you routinely shop, but we must do it, so let's start with baby steps. Here are a few easy changes you can make to your buying habits that will help reduce soft plastics.

- Buy products in bulk. Buying larger packages reduces the amount of plastic used.
- Shop at grocers that offer bulk food options for things like baking goods, cereal, and snacks. If they don't allow you to use your own container, ask if they have paper bags instead of plastic.
  - I use my mesh produce bags for bulk food buys like nuts.
  - In some communities, grocers are required to allow shoppers to use their own containers.
  - If you use a plastic bag for bulk food when it's empty, store it with your reusable bags for your next trip to the bulk food aisle.
- Buy from a local butcher or a grocer with a meat counter. Ask them to wrap your meat in coated paper instead of plastic.

- Though this coated paper can't be recycled, it's less damaging to the planet than the plastic and foam used for pre-packaged meats.

Tip 10 - Plastic food wraps and baggies can't be recycled. Reduce and reuse these.

There are alternatives to zip-close disposable baggies and plastic wraps we use to store food, and they're a lot healthier for us and the planet. So switch to these reusable options.

- Glass or plastic reusable food containers are cheaper since buying baggies, and plastic wraps can quickly add.
- There are reusable zip-close food storage bags. These washable bags are a treated plastic and can even go in the dishwasher.
- Wash and reuse disposable food storage baggies.
  - One box of zip baggies lasted me over a year. I've washed and reused them until they're worn out or too soiled.
- If you use plastic wrap for storing leftovers, finish up what you have and stop buying it. This type of soft plastic is not recyclable.
  - Wash and reuse plastic wraps when they aren't too messy.
  - Use aluminum foil or wax paper to wrap and store food.
- Make your own reusable beeswax food wraps for wrapping lunches and leftovers.
  - They're easy to make; look for instructions online.
- Check out the Japanese art of furoshiki, a type of reusable cloth wrap. There's instruction online on how to make these.
  - What a great way to cut up and reuse old clothing that can't be donated.

Give up the convenience of packing lunches in throw-away containers and build the habit of bringing home your reusable ones—what a great way to teach your kids responsibility.

Polystyrene foam - a recycling nightmare and planet killer.

The worst plastic we use in abundance is polystyrene foam, aka StyrofoamTM.

Recycling polystyrene foam into something else is difficult. While there are special handling facilities that do it, this toxic material is used so much in our world most of it won't get recycled. Plus, there isn't a big market for recycled foam since making new is so cheap.

Tip 11 - Polystyrene foam (StyrofoamTM) can't go in your recycle bin.

I see it all the time, StyrofoamTM in recycling bins where it doesn't belong. Even if there's a recycling triangle on it, this foam is a soft plastic that can't go in your bin. It comes apart easily and clings to everything causing contamination of items that otherwise could have been recycled.

I learned this as a recycling volunteer at a town where they accepted polystyrene at a drop-off-only program.

These programs are rare because recycling this material isn't easy, and there aren't many buyers for it.

If you think your recycling service has a drop-off program, give them a call, or check their website. If they don't take it, they may know a place that does.

Check the Earth911.com database to find a drop-off site in your area.

Tip 12 - Ways to eliminate toxins in your food from polystyrene foam.

Like any plastic, polystyrene foam is made from toxic petrochemicals. In addition, because it is a loose soft plastic, this foam can easily break down and get in your food.

It's hard to eliminate foam food containers when you dine out or carry out, but this soft plastic can break down or melt and get into your food. Though fast-food restaurants use them less now, foam containers are the cheapest to-go boxes restaurants can offer for take-out and leftovers, so they're still the most common.

- Since you should not waste food, take home your leftovers and eat them but find ways to avoid foam containers.
    - BYOC - Bring your own container or ask for an alternative to foam.
        - Buy reusable zip bags that are easy to carry, put your leftovers in, and... wash, rinse, repeat.
- Learn more about how restaurants are eliminating unhealthy containers at Green Restaurant Association, dinegreen.com.
- Foam cups are often used for coffee and hot beverages. Carry your own travel mug to avoid these. Don't have a travel mug? Go for the paper cup. It may be coated with plastic and not recyclable, but these aren't as damaging as foam.

Polystyrene foam and plastic packaging were never intended to be anything more than single-use. However, since these aren't easy to recycle, it's important we find ways to stop using them.

Dispose of Your Plastics Habit - How to Start Today

Over half of our single-use plastics were created in the past 15 years. They became a lightweight and convenient alternative, and they're a cheaper form of packaging that manufacturers prefer.

Tip 13 - How to reduce foam and plastic packaging in products and shipments.

It's fun to get a new TV or appliance, but with it comes a lot of packaging. Though it's hard to say no to buying something you need, like a new appliance, letting manufacturers know that you want to buy products with greener packaging helps.

- Look for sustainable packaging options on the retailer's website.
- When you place an online order, add a comment to tell them you prefer sustainable or plastic-free packaging.
- Submit customer comments to companies and let them know you want them to offer sustainable products and packaging.
  - If enough of us speak up, manufacturers will pay attention.

See **Tip 51** for ways to reuse foam packing materials.

Tip 14 - We need to stop using 3 - 7's and other non-recyclable plastics.

Since 1 and 2 plastics are the only ones that are likely to get recycled, it's important that we reduce our use of the others, even if they have the recycling triangle and your MRF says they take them.

There is no way to recycle most plastics.

In the 70s and 80s, plastic trash started showing up all over, and people began to complain. Anti-plastic campaigns were started, and grassroots organizations came together to start events like Earth Day. The future of disposable plastics looked grim.

- This durable material turned out to be not so great for the planet since it's so durable it doesn't dissolve in a few years like most trash.
- Plastics stick around for hundreds of years.

The fossil fuel industry invested millions in an anti-litter campaign called *Keep America Beautiful*. It kept people distracted from thinking about what happened to waste after we disposed of it. As a result, litter in our neighborhoods went down, and this *out-of-sight, out-of-mind* strategy seemed to have fixed their market problem. Until it didn't.

To keep selling plastics, the fossil fuel industry had to make people believe they could be recycled. But there's only so much space on

Earth, so eventually, plastics became a problem, and we figured out they weren't being recycled.

- A lot of psychology goes into marketing, and the fossil fuel industry used it to convince us that plastic recycling worked. We just had to do it right. This was BS.
- The use of the recycling triangle and actions of the fossil fuel industry have been likened to those of tobacco companies and how they hid what they knew about the damage their products cause.

Tips to de-plasticize your life

If you want to make a positive impact on the life of our planet and future generations, paying attention to plastics and reducing them is the best place to start.

Tip 15 - Don't buy bottled water. It's not better for you, and it's bad for the planet.

Over 50 billion water bottles are sold each year, and only about 20% of those get recycled.

A few decades ago, people didn't walk out of stores with cases of bottled water. Now I see it all the time. It's not better water, so what people buy is the convenience. Even if the empties are put in recycling, there is a slim chance they will get recycled.

- Studies have shown that most bottled water is no better for you than tap water.
  - There are exceptions where tap water is contaminated by lead pipes, like in Flint, Michigan. But most municipal water systems do a good job of filtering and treating water, so it's usually better for you than bottled.
- For the purest water, buy a filtration pitcher or purifier and carry a reusable water bottle. This is the healthiest way to drink water, for you and the planet.

Tip 16 - Say no thanks to plastic straws. Get a reusable one.

In America, over 500 million plastic straws are used every day. This is especially bad because straws cannot be recycled. They are soft plastic that can only go in the trash.

- When dining out, be proactive and hand the straw back to your waitperson. If you don't use it and leave it on the table, it will likely end up in the trash.
  - The decision not to use a straw is the first step; making sure it doesn't end up in the garbage is equally important. Hand it back and say, *no thanks*.
- Straws are another reason to minimize your fast-food habit. It's hard to avoid if you enjoy road trips, but carrying a water bottle and packing food and drinks can help.
  - If you go to a drive-thru while on a road trip, save the first straw to use later or carry a reusable one.
- Buy reusable straws in stores or online.
  - The portable ones I have are collapsible and come in a little holder so I can carry them in my purse.
- If you dine inside a fast-food restaurant, skip the lid and straw. Neither of these is recyclable, even if there's a recycling triangle on the lid.

Straws may be a tiny thing, but we use over 180 billion a year. That's just in the US. These small non-recyclable plastics often end up in our oceans, harming sea life.

Tip 17 - Refuse plastic utensils when offered, and don't buy them.

Plastic eating utensils are not recyclable, so don't buy them, even for events.

- If you're having guests, pick up utensils at a thrift store. You can donate them later.
- Companies with rental equipment for weddings, graduations, etc., also rent dishes and utensils.

- Just Say No! to the utensils they try and give you with your carry-out meal.
  - If you're heading home, you have utensils there.
  - Not heading home? Carry them in your car and keep some at work.
    - Pick some up at a thrift store, so you don't lose your home dishes.
    - Buy and carry portable utensils.

Back when I ate fast food more often, I hated it when I'd hit a drive-thru on my way home, and they would throw plastic utensils in with my salad. I wouldn't use them since I ate at home; instead, I kept them for camping. I wash and reuse them until they break.

- When I hit a drive-thru now, I ask for no utensils. They still sometimes end up in my bag, and I continue to keep them for camping and road trips, but when they leave them out, it feels like a win.

Tip 18 - Don't buy bubble wrap when you move; there are free alternatives.

When it's time to move, don't buy disposable packing materials like bubble wrap. You already have what you need to keep your breakables safe.

- Wrap dishes and other breakables in clothes, towels, and sheets.
- Use shredded paper and packing materials you save from shipments you receive.
- Cut cardboard into spacers if you need to secure breakables like glasses.
- As your move approaches, keep an eye on your recycling for items you can reuse as padding or separators, like cereal boxes and egg cartons.

Packaging and the not-so-joyful part of online shopping

The Covid pandemic saw a huge boost to online shopping as people avoided leaving their homes. Over half of the people asked in a survey said their online shopping had increased since the pandemic started. This caused a big increase in packaging use.

Be a greener online shopper - your wallet and planet will thank you.

Tip 19 - Consolidate your online orders to reduce packaging and emissions.

Online shopping can be easy, efficient, fun, and addicting. Everybody likes to get new stuff. It's part of being human. But most of it comes with plastic packaging that is harmful to the planet. Changing how you shop can reduce this impact.

It's handy that we can pick up our smartphones and, within minutes, buy something that will show up on our doorstep. I thought this was helpful until I realized I didn't need most of it right away, if at all.

- When you don't need something immediately, write it down. Once you have a list of items, order them all at once.
- Add items to your shopping cart when you think of them, but don't place the order until you need to.

Making this simple adjustment will help you rethink the need to buy something, saving you money and helping to consolidate shipments and reduce packaging.

- When shopping online, always put in your order comments that you would prefer plastic-free packaging. Companies will fulfill your request if they can.

It's easier to change shopping habits when you make a conscious effort.

The Impact of Plastics - The Good, Bad, and Very Ugly

Plastics have an interesting history that ties to how unprepared we were for our modern waste habits. When it became a mainstream commodity in the 1950s, it was considered a miracle material that could solve a lot of our problems, and it did. We benefited from plastics and liked the fact that we could easily dispose of them.

Where would we be without plastics? A necessity in today's world

There are several benefits to plastics.

- Football players used to wear leather helmets before they had the protection of plastic ones.
- In medicine, disposable plastic syringes and other treatment materials protect us from infection. Plastics are a critical material in prosthetics.

Plastics have benefits in transportation and technology.

- Today most vehicles and other forms of transportation incorporate plastics to make them lighter weight, more fuel efficient, and safer.
- Every electronic device you own has some plastics in it.

So, yeah, plastics are important. However, what we need less of are single-use plastics.

Single-use plastics - a convenience or a death trap

Though there are a lot of benefits to plastics, we need to stop looking at this toxic material as disposable. To dispose of something means it will go away. But unfortunately, plastics won't go away – we're stuck with them.

The most dangerous thing about plastics is that they can take hundreds or even thousands of years to dissolve, depending on the type. So instead, they break down into tiny pieces and microplastics that show up on our beaches and at alarming rates in fish and other animals, including humans.

Tip 20 - Avoid pre-packaged meal delivery services that aren't eco-friendly.

Meal kit programs that ship you pre-planned recipes and ingredients are a great time saver. They cut out meal prep and planning time but can be costly to the planet.

I tried this once and was blown away by all the plastic packaging. Each pre-measured item was individually packaged in plastic. I get that this keeps ingredients fresh, but if I do this again, I will go with an eco-friendly company that uses recycled or biodegradable packaging.

- Look for meal kit programs that use eco-friendly packaging, as well as organic and responsibly sourced ingredients.

Yes, it's a hassle to change how we buy things, but the convenience of living a disposable lifestyle is catching up to us. We need to make changes now, or life will become much less convenient.

How and where plastics show up in our lives

At first, I didn't think about the increase in plastics produced during the Covid-19 pandemic, maybe because it played an important role in keeping us safe.

- Plastics are a strong material used extensively in healthcare.
- They're used to make disposable masks, face shields, and plexiglass.
- None of these can be recycled.

The plexiglass put up in stores, banks, restaurants, or any business that wanted to stay open will end up in landfills when they come down. One University Professor challenged students to come up with ways to reuse these. Unfortunately, I've heard of no solutions yet.

Manufacturing of plastics is expected to grow by over 30% by 2030.

All this emergency use of plastics has piled on top of what we already use, and at a time when we're facing a crisis from having too much of it. So much that it's harming us.

Tip 21 - How plastics show up in our food and affect our health.

Microplastics are now found in animals. They're being detected in things like honey and table salt and are found in food crops, ocean life, and even human organs. They're so tiny they can even end up in our drinking water.

It is estimated that each of us will consume around 40 pounds of plastics in our lifetime.

Chemicals in plastics have been proven to cause cancer, infertility, and brain disorders, as well as a slew of other health problems. With all the plastics in the world, it's not hard to figure out that these toxins are affecting our health.

- This is only the damage done when we dispose of our plastics.
- Emissions from making plastics will exceed the coal industry within ten years if demand continues to increase.

Controlling Our Plastic Use - A Matter of Survival

Because there's so much of it, the environmental impact of making and disposing of plastics is quite serious.

- Manufacturing plastics is energy-intensive and accounts for around 10% of our fossil fuel use.

A study conducted on 2015 emissions concluded that, in the US alone, annual plastics production created around ten million tons of carbon dioxide.

- This equals the emissions from driving two million cars for a year.
- This is only from *making* plastics; it doesn't account for emissions caused by transporting and disposing of them.

As plastics production grew, how we dispose of it became a challenge. Wealthier consumerist nations began shipping it to developing countries that didn't have the infrastructure to properly dispose of it.

The Basel Convention is an intergovernmental organization that regulates the type of waste that countries can ship to other countries, in particular hazardous waste.

- In 2019, they added plastics to the list of hazardous materials not allowed. Over 180 countries agreed to ratify it.
  - Not all member countries, including the US, have ratified this new regulation. But some US states are moving forward with similar laws.

Single-use plastics are so damaging to our environment that in May of 2021, the US Department of Energy announced they are investing $14.5 million in research on how to reduce them.

We're drowning in plastics - it has to stop!

If you've ever watched the Disney movie WALL-E, you may have felt the same dread I did when it showed Earth covered in garbage, our planet uninhabitable. Though it's a delightful movie, it also gives a stark realization of what the world will look like if we don't change. I've seen many post-apocalyptic movies, and some seem like they could happen. However, it feels like the world of WALL-E is already happening.

If we're going to stop the flood of plastics polluting our oceans and planet, three things need to happen right away.

- Stronger legislation that requires companies to use recycled packaging and take responsibility for what happens to single-use plastics after we're done with them.
    - We must require manufacturers to package products in plastics that can be recycled.
- Reduce our demand for plastics. Something we can do with our habits, voice, and wallet. Work on your list from **Tip 2**.
- Everyone buying recycled packaging to increase the demand and lower prices.
    - This will make recycled plastics more affordable and easier for companies to use in products and packaging.

The recycling of plastics may be failing us, but unless we plan to eliminate disposable packaging, it is the only solution we have. So we need to get this right.

Tip 22 - Vote locally for laws to reduce plastics and what's allowed in your community.

Laws and regulations for recycling and materials allowed in your community are determined by state and local governments and what you support.

- Many local laws happen because politicians are pressured by their constituents.
  - If they don't hear from the people they serve, they listen to corporate lobbyists.
    - It was plastics industry lobbyists who pushed for laws in US states that prohibit bans on single-use plastics and polystyrene foam.
  - Bans against bans, or preemptive laws, are not bills you'll see on your voting ballot. So, it's critical we elect representatives who support environmentally friendly initiatives and laws.

Colorado, a US state with laws preventing bans on plastics, had many communities that didn't follow the anti-ban law.

- People came together to push for local laws to fight single-use plastics.
  - Many communities started writing bans on plastic shopping bags and take-out containers.
- Due to pressure from Colorado residents, the governor signed a law to ban plastic shopping bags and foam take-out containers. It goes into effect in 2024.

Local governments and community grassroots efforts make the biggest impact in effecting change.

"Never doubt that a small group of thoughtful committed citizens
can change the world. Indeed, it is the only thing that ever has."
Margaret Mead (1901-1978), American cultural anthropologist

A plastics war - hope for the future

No one is oblivious to our plastics problem. Many governments are looking at how to solve it.

- In the UK, the Prime Minister asked grocers to provide bulk food aisles, and in 2022 they started taxing plastic packaging that is not made with at least 30% recycled material.
  - Some US states have similar laws.
- Many countries have Extended Producer Responsibility (EPR) laws which require companies to take full responsibility for the environmental impact of their packaging and include making sure it can be recycled.
  - States in the US are beginning to enact EPR laws.
- Laws like these, and the Break Free from Plastic Pollution Act being pushed for in the US, are long uphill battles.
  - They take a long time to pass.
  - They need our support if they're going to happen and be effective.

Speaking up and voting is the only way to get laws that will protect us from corporate polluters and the plastics industry.

Grassroots efforts are the catalyst for changes to local laws.

Our growing plastics waste has created a supply of materials now being used for a multitude of things around the globe.

- Because it's such a durable product, in Germany, recycled plastics and old tires are being used for the repair and new construction of roads.
- In India, recycled plastics are used to make cooking oil for households.
- In the US, plastics are being used in the production of jet and diesel fuels.
- Clothing companies are starting to use recycled plastics to make polyester.

- o Imagine the difference it would make if all our clothes were made from recycled materials.

These are just a few of the opportunities that bring jobs and grow the economy while cleaning up plastics. But we have a whole lot of plastics to deal with, so we need a lot more of these actions.

Tackling our Plastics Use Will Put a Big Dent in Our Waste Problem

Cleaning up our plastics waste and slowing down the demand for new plastics will take a global effort. From governments and industries all the way down to you and me. We're all in this together.

Tip 23 - Ways to support efforts to reduce single-use plastics.

We need to educate ourselves and support organizations fighting to stop new plastics and eliminate them in our environment. Grassroots and non-profit organizations are good sources.

- PlasticOcean.org works with the UN's Sustainable Development Goals to support plastic clean-up of coastal communities around the globe.
- Earth Day Plastic Action is sponsored by EarthDay.org and teaches 6Rs, adding Remove, Refuse, and Rally to the cry to control our plastics.
- PlasticBagLaws.org is a resource for current legislation that limits the use of plastic bags. A good source for legislators looking to enact bans in their district.
  - o Why not email your local representatives and share the website with them? Maybe they're looking for something like this; they will be if enough of their voters tell them to.
- BeyondPlastics.net - helping us learn about successful projects that are working to replace plastic packaging and more.

- Natural Resources Defense Council (NRDC) provides information on single-use plastics and their damage to the planet. It can be found at nrdc.org.
- National Geographic has a series called *Plastic or Planet* which can be found on their site at nationalgeographic.com.

Make sure you verify the information you read against scientific data. Remember, as much good information as there is on the internet, there is equally misleading information.

We need to make fewer plastics, not more

Since plastics recycling hasn't worked, we would have been better off if we had focused on reducing and reusing. Now that we know we can't sort our way out of the problem, reducing and eliminating plastics is our only hope.

- Most other disposables can be recycled, so waste *will* go down if we focus on reducing single-use plastics and reusing what we have.

We can't get rid of plastics, they're everywhere, but if we change how we look at them, we may be able to get them under control. Let's start by looking at plastics as the durable material they were intended to be and not disposable.

- What's ironic is plastics may have made our lives easier for the past fifty years, but they'll make our lives much harder for the next fifty and beyond.

Start with these main ideas to get your plastics consumption under control. Then keep adding more habits from tips in this and other chapters.

1. Make a list of where you find plastics and StyrofoamTM in your life. Work to reduce and eliminate them.

2. Get reusable shopping bags, buy in bulk, and limit or get rid of carry-out and drive-thru eating habits.
3. Always properly recycle plastics, and only what you know may get recycled.
   a. Check with your service and only put the plastics they take in your bin.
4. Avoid plastics that don't recycle. Plastics with the triangle that says 6 or 7 can't be recycled, and 3's and 4's are soft plastics that rarely will.
   a. Reuse disposable plastics for as long as you can. You will save money and the planet.
5. Buy recycled packaging whenever possible. Our demand for post-consumer materials helps plastics recycling work how it was meant to.
   a. Support better recycling laws to reduce new plastics and increase the use of recycled.
   b. This is how we build a circular economy with our disposables.
6. Extended Producer Responsibility (EPR) laws are needed to hold manufacturers accountable for the disposable plastics they make and sell.

Chapter 2: Reduce is the Only Way to Stop Waste from Growing

The problem with disposables is that they're so darn convenient. The technology exists to eliminate them, and we have convenient reusable solutions available. We must change our habits and start using these.

To clean up the planet, we must dispose of our disposable lifestyle.

Would it really be that hard to carry home a reusable food storage bag to toss in the dishwasher after taking a sandwich to work? Or cut back on carry-out eating?

Because we've gotten used to a disposable lifestyle, reducing waste is one of the biggest challenges we face.

It's going to take each of us thinking about how we buy things and the companies making them so they last longer and can be repaired. It's going to take our wallets and our voices to push for reduced and responsible packaging.

The First of the 3Rs for a Reason - Reducing Waste is Critical

Studies show that developed countries generate three times more waste than underdeveloped countries.

- The amount of trash generated in the US each day averages out to about five pounds per person.
- Over 80% of goods purchased are thrown out within a year.

Tip 24 - Don't just toss it; think about it. Pay attention to how much trash you generate.

It can be illuminating to look at everything you throw out with curiosity. Stop and ask yourself, "Why did I generate this trash?" Was it because you drank a soda or forgot your lunch and picked up take-out?

- Keep cans of soup, a microwavable bowl, and a spoon at work.
    - Soup cans recycle, but carry-out containers don't.

Whatever the reason, give your trash some thought. This includes what goes into recycling.

- Did you need to do whatever it was that caused you to generate it?
- How can you change that practice?
    - Get creative. Talk to your family and make a plan.
- Did you need to print that email or buy that bottle of water? Even if you recycle these, it's still waste that you have generated.
- Pick one piece of packaging you regularly buy and find ways to eliminate it.

Start today with these tips to reduce waste

The best way to reduce your waste is to start with a few simple changes and continue from there. For example, pick one or two things in this chapter you can do now, then add more and keep adding.

Most importantly, act! Don't think about how hard or inconvenient it might be. The more you think about it, the less likely you are to do it.

If something feels too hard to implement, keep looking until you find changes that work for you. These are lifestyle changes, so they're going to take time. Don't get frustrated. Keep going. This is important.

How we use and waste our resources has huge consequences.

- Many we never see or know about.
- Some are bigger than others.

The waste we don't see, what governments and corporations want to hide from us, we'll talk about later. But first, let's look at the waste we can control, namely the trash we generate and how to cut down on it.

In your home and life

Tip 25 - How to reduce or eliminate using paper products in your home.

Half-a-million trees are cut down every day. Since trees naturally sequester carbon dioxide (CO_2), which we need to reverse global warming, saving trees is important. While not all trees are cut down to make paper products, many are. Most are for convenience.

There are several paper products that have reusable alternatives. It takes some effort, but it's worth it.

- Don't buy paper towels - For everything you use paper towels on, like wiping up spills, there's a reusable option available.
    - In the US, people use more paper towels than all the rest of the world combined. They're just too darn convenient.
- Use cloth dinner napkins - As I started reducing waste, I stopped buying paper napkins but found I used paper towels instead. Unfortunately, neither can be recycled, so I started looking in thrift stores and yard sales for cloth napkins I wash and reuse.

- Use a sponge to clean up spills - Some people worry about the germs a sponge carries, but most cleaners you use on it have disinfectant.
  - Toss your kitchen sponge in the dishwasher. The hot water and dish soap will kill the germs.
- Use newspaper or old rags to clean mirrors and windows - When I donate clothes, I always find a few items that are too old and ratty. So I cut them up into cleaning rags. I've also bought bundles of rags at thrift stores.
- Buy toilet paper made from recycled paper or bamboo - Online retailers who sell these offer bulk options and use sustainable packaging.
- Reduce your toilet paper use - Pay attention to how much you use vs. how much you need. Roll back the roll to use only what you need and avoid waste.
  - Get a bidet. You can get a portable one for around $100. You'll spend less on toilet paper.
- Go old school and use a hanky on your nose instead of a tissue.
- Don't buy new books - read digital books, buy used ones, or visit your library.
  - Some libraries loan out more than books, like sewing machines and tools.
- Reuse printer paper that's blank on one side - Reuse the blank side for scratch paper and notes.

Tip 26 – Opt out of unwanted junk mail to reduce your use of trees.

- It is estimated that junk mail in the US takes out 100 million trees each year and uses as much energy as three million cars.

Cutting down a tree releases CO_2 into our environment. More CO_2 is emitted when the wood is milled into paper, and printing begins. Finally, there are emissions from processing and preparing the mailing piece and the transportation to get it to your mailbox where... you usually toss it out.

That's a lot of emissions from something that ends up in the recycle bin, where it continues to increase carbon emissions as it's hauled away and hopefully recycled.

- If you read your weekly flyers or junk mail, great. It wasn't a waste getting them to you. But if you don't read them, there are ways to eliminate them.

About 80% of direct mail companies belong to DMAChoice.org, where, for a $2 processing fee, you can opt out of new mailing lists for ten years. It also provides a free way to opt out of credit card offers. I don't know about you, but I find it much healthier for my finances not to be tempted by those.

If you receive coupon mailers like Valpak or Red Plum that you never look at, go to their website and opt out.

- For Red Plum, visit save.com, scroll to the bottom, and click on Delivery Options.
- For Valpak mailings, go to valpak.com/remove-address.

These companies would rather not spend the money mailing you something you don't want.

Tip 27 - Avoid pre-packaged foods for a healthier you and planet.

Find alternatives to excess packaging when you buy food. There are companies that offer sustainable products and buying options, from local markets to online retailers.

Reduce waste and emissions by eating locally grown and whole foods.

Individually wrapped snack foods are convenient, but it doesn't take a lot of effort to cut them out.

- Don't buy single-serve or individually wrapped foods. If you enjoy vegetables or chips with your lunch, buy fresh produce and a big bag of chips, then use reusable baggies or containers for lunches or travel.

- Look at anything you eat that comes individually wrapped and think of alternatives. I'm sure you can come up with great ideas.

Tip 28 - Find ways to eliminate single-use beverage containers from your life.

A lot of beverages come in single-use containers, from soda to energy drinks, juices, and vitamin water, the list goes on. So pay attention to how many you use in a day and find ways to reduce this disposable habit.

- Cut out the energy drinks and switch to coffee or tea you can carry in a reusable travel mug.
  - o An exercise routine in the morning can give you more energy.
- Buy soda in two-liter bottles or get a soda maker.
- Keep pitchers of beverages like Kool-Aid or juice in the fridge.
- Beer and wine can be purchased in big containers.
- Never buy bottled water. Instead, get a water filter and a reusable bottle.

Tip 29 - Eliminate carry-out and fast food to save money and the planet.

The best thing we can do is cut carry-out food containers completely from our lives, but for many people, a more realistic approach is to cut back.

In today's busy world, we often default to drive-thru or take-out for dinner because we run out of time to prepare food. It's understandable. If you do this often, find ways to cut back. It's good for you, your budget, and the planet.

For one week, keep track of how much you eat fast food or carry out. If it's only once or twice a week, congratulations! That's less than most people.

The average American eats out four to five times a week. Three of those are visits to fast-food restaurants.

- Commit to better meal planning. Over the weekend, write a meal plan for the week.
    - Look online for meals you can put together in advance or take very little prep time.
    - Having something ready to go in the oven will make it easier to skip a drive-thru since you won't want to waste the food you have already prepared.
- Make it a family activity. Kids love to help in the kitchen. Get them involved with preparing meals for the week.
    - This is a good time for family conversation. You could talk about reducing waste and why it's important.
        - *Build these habits together.*
- Put leftovers into reusable containers, ready to take for lunches.
- Look at your bank account and add how much you spend eating out each week. Monitoring your spending can discourage this expensive habit.
    - You will quickly realize how much you will save cooking at home and carrying lunches.
- Count how many pieces of trash come from one trip through the drive-thru.
    - None of it can be recycled.
        - Clamshells may have a triangle, but these plastic and foam containers won't recycle.
- Look for an eco-friendly restaurant to dine at or pick up food. Check out the Green Restaurant Association, dinegreen.com, or Food Print, foodprint.org/dining-out-sustainably, to learn how.

Tip 30 - Think before you print and encourage others to do the same.

Recycling a single piece of paper seems so minor, but the average office worker will use 10,000 a year. That's a lot of trees we can save if we reduce our paper use.

- At home or work, think about why you print a document.

- Set your default to two-sided printing for those times you do have to print.
- Encourage others to think before they print.
  - Add "Be earth-friendly, think before you print" to your email signature. Little things like this help with awareness.

At your workplace

I've worked for big and small companies, and I've found the attitude toward sustainability means everything. For example, I started a job once at a company with about fifty employees. There was no recycling in the kitchens, so I took mine home. Then one day, I asked why we didn't have it. The answer was simple; no one had ever started it. So I found some boxes, a corner in the kitchens for them, put signs on them, and got approval to put them out.

- I had to gather the recycling weekly and take it to a drop-off, but eventually, my coworkers began to help, and the company started to pay the $3 drop-off fee.
  - Soon after, I started the office green team. We did community clean-up events and helped employees learn how to recycle and reduce waste. They now have an annual budget and real recycling bins.

It wasn't hard to get people on board and make recycling a practice at my workplace. Most people want to help build a healthier planet, but they don't always know how.

Tip 31 - Talk with coworkers and encourage waste reduction at work.

Strike up a conversation with coworkers about recycling and how to reduce waste, or talk about a documentary or video you watched recently about the environment.

- They may know of other educational materials or have tips you haven't heard.

- Be curious about other people's actions and habits. Maybe you will both get excited about an environmental topic and spearhead change.
- Don't get discouraged when people talk about how recycling doesn't work.
  - This is not a reason to keep creating excessive waste or stop recycling.

Warning: Much of what we hear in conversation is hearsay or outdated. Before you act on it or repeat it, look for the science and data to verify it. This is a great way to get educated and learn what really makes a difference.

Talk to your manager or HR department about sustainable practices at work.

- Speak up if you notice changes are needed, like better recycling or eliminating StyrofoamTM coffee cups.
- Have you ever noticed how office buildings have lights on all night? I have, and it bugs the heck out of me. Ask if your office lights are on a timer, so they shut off overnight. If the answer is no, ask why not.
  - Installing technology that shuts down everything except security lights may have a price tag, but the energy savings will pay for it.
- If your workplace doesn't have light sensors in bathrooms and conference rooms, ask if they will install them.
  - Newer office buildings may include these because sustainable building codes require them.

Tip 32 - Find ways to reduce business trips and hold remote meetings whenever possible.

Many companies are embracing remote meetings, especially since the Covid pandemic started. Ask if flying to a meeting is necessary. Hopefully, your boss will support your efforts and will appreciate you trying to save money and be more sustainable.

- If your company has a policy in place to reduce emissions, cutting back on air travel can boost the sustainability profile and public relations.
- Meetings can be slowed down by technical difficulties when done virtually. The time saved by not traveling to those meetings can make up for it.

Tip 33 - Reduce your environmental impact at work from chemicals and emissions.

If you work with chemicals, there are laws that regulate their use and disposal. Be mindful to follow them correctly and encourage others to do the same. These laws are in place to protect people and the environment.

- Go above and beyond the regulations. Investigate the chemicals you work with and see if there are fewer toxic alternatives. Suggest these to your company. They may not know there are alternatives.
    - They could be using toxic chemicals because that's the way it's always been done.
    - We need to speak up to fix our systems. Taking the initiative at work is one way you can do that.

A lot of vehicles are regularly left idling. Semi-trucks do this all the time. If you work in the trucking industry, ask why this is a practice. It may be company policy, or it may be because that's just the way it's always been done. Either way, help your company and coworkers change this mindset.

- Idling may be necessary for extreme weather so drivers can sleep and stay healthy, but I've driven through many US states and saw idling semi-trucks in rest areas when temperatures were mild.
- Fortunately, delivery drivers now shut down their vehicles while parked. Though this may be to prevent theft, it is helping to reduce emissions.

More efforts need to be made if we're going to reduce emissions from transportation, the biggest cause of greenhouse gases. It will likely take regulations that get more electric vehicles on the roads, but in the meantime preventing unnecessary idling is a step in the right direction.

Zero Waste is Doable - Tips on How to Get Started

Our waste has become very costly. In the US, most communities spend more on waste management than on schools, fire protection, libraries, and parks **combined**.

The idea of living a zero-waste life is not new; in fact, it's pretty darn old. But nowadays, with all the disposable stuff in our world, it seems like a daunting task. It's not!

- I was blown away when I read about a zero-waste aficionado who fit all her trash from a two-year period into a 16-ounce mason jar.
    - This should be the goal of every one of us. Sound hard? Not really. It is a process that will take work, but it's not hard if you don't let up and keep making changes.

Baby steps to zero waste in your home and community.

As you incorporate the habits in this book, you will start seeing a reduction in your waste. This is a good time to set zero-waste goals.

Tip 34 - Revamp your shopping habits to eliminate non-recyclable packaging.

Once you start paying attention to your waste, you will notice that a lot of packaging is not recyclable.

- Look at where packaging comes from and find ways to stop buying it. Some will be easy, like buying your produce loose instead of pre-packaged, while others may take more effort.
- Shop at grocery stores that offer bulk foods and ask if you can use your own container or reusable bag.
  - Do they offer paper bags or containers?
  - Save the bag or container from your last visit and reuse it.
- Buy from specialty shops and local family-owned stores that offer less or no packaging.
- Avoid plastic containers. Living a zero-waste life means using only packaging that can be reused or recycled.

Tip 35 - Where to find zero waste and sustainable shopping options online.

The global effort to eliminate single-use packaging and provide sustainable products is growing, so there are many opportunities to buy zero-waste products.

Here are online retailers that offer sustainable and ethical products. Many have *give-back* programs where your purchases also support people in need.

- Simple Switch, simpleswitch.org, offers a way to buy sustainable goods while empowering people in need of support.
- Earth Hero, earthhero.com, has ethically sourced products and offers zero-waste shopping on their site.
- Zero Waste Club, z-w-c.com, offers sustainable products like toothbrushes, razors, collapsible reuse straws, and other commonly used items.
- Thrive Market, thrivemarket.com, has a full range of groceries sourced and packaged responsibly and sustainably.
- Done Good, donegood.co, is a centralized marketplace for finding ethically and sustainably sourced products.

There are always new companies offering sustainable buying choices. Search the internet to find the ones that fit your needs.

Think of Mother Earth when planning your next get together.

A little planning can change your get-togethers from creating bags and bags of garbage to generating little or no waste.

Tip 36 - Go zero waste when entertaining, have reusables, and set up recycle bins.

When you throw a party or have company over, it's easy to default to disposable conveniences. While this may make cleaning easier, it has environmental consequences that can be avoided.

- Buy reusable utensils and plates at a garage sale or thrift store.
    - Donate them afterward if you don't want to keep them or have the space.
    - You can also rent dishes from a party rental place.
- Beware of compostable utensils. This labeling is not regulated. You must investigate the claim before you compost them.
    - If you don't have composting available, don't waste your money. They will have to go in your garbage. These can't go into the recycling.

Compostable doesn't mean recyclable.

- For beverages, make pitchers of tea and lemonade or buy big containers like two liters of soda.
    - If you prefer to fill a cooler with cans and bottles, stick to aluminum and glass; they're more likely to get recycled.
- Buy reusable cups you can pick up used or at a dollar store. Get them in multiple colors and styles so your guests can tell which cup is theirs.
    - Disposable beverage cups are not recyclable, even plastic ones with a recycling triangle.

- Plain white napkins with no inks or dyes are compostable. If you don't have compost, these go in the trash and are not recyclable.
  - Consider buying or making cloth napkins to offer your guests. If they're eco-minded, they will appreciate the effort.

Save a few cardboard boxes, mark them as recycling, and point them out to your guests.

- Always rinse cans and bottles before you put them in your recycling bin or haul them to a drop-off.

It feels good to know your event won't make a big impact on the planet. Who knows, you may influence others to do the same.

Tip 37 - Washing dishes isn't as bad for the environment as using disposables.

Using regular dishes instead of disposable ones reduces your impact when done right.

While some people may think it's better not to use energy and water to wash dishes, using disposables is much worse. The making and disposing of single-use items wastes a lot of energy and water.

- Most dishwashers are energy efficient and use less water.
- Load your dishwasher fully and correctly and set it to air dry.
- Skip the pre-wash cycle. It's only needed if food is stuck to dishes; wipe them off before loading them in the dishwasher.

If you don't have a dishwasher, the most efficient way to hand wash dishes is in a tub or a sink filled with soapy water. Never wash with the faucet running.

- Hand washing dishes with the water constantly running can use twice as much water as a dishwasher.
  - Hand washing in a tub or sink of soapy water uses about 30% fewer emissions and significantly less water.

o   Rinse dishes with the faucet on low stream.

Visit goingzerowaste.com and subscribe to the newsletter to learn more about how to eliminate waste and live sustainably.

Waste and the Damage It Causes

While reducing our trash is critical, how much we waste our resources also makes an enormous impact on the environment.

Tip 38 - Cut out food waste; it's more damaging than you may realize.

Food waste is something we have direct control over. Food production uses a lot of energy, water, and threatened natural resources. The environmental impacts of livestock and meat production are so significant they've become major contributors to our unhealthy environment.

One-third of food produced is wasted and ends up in landfills.

It's hard to understand why over three million children die in our world every year from starvation while a third of our food and half of the produce grown is thrown away.

- Food waste in the US costs us over $160 billion annually. That's about $1,500 per household.
- Nearly 10% of our energy goes toward producing food. Since a third of that is wasted, we could reduce over 3% of our energy use by eliminating food waste.
- Food production uses around 60 million gallons of water per second. That's over 60% of our water use, a third of which could be avoided.

We have control over how much food we waste. While not all of it is from individuals, much of it is. Let's work together to eliminate food waste and the damage it causes.

- Make a grocery list and stick to it. This is the best way to not buy extra food that may not get eaten.
- Eat the whole fruit or veggie when you can, including the skins.
  - Most vitamins and nutrients are in the skin.
- Use citrus peels to flavor your water, so you enjoy drinking more of it.
- When cooking a meal, make extra to have leftovers for lunches or another meal.
  - Don't throw out uneaten food. Instead, save it for your next meal or snack.
- Only put on your plate what you know you will eat.

If you find you throw out food that went bad in your fridge, think about the foods you toss and how to avoid this action.

- Buy foods that go bad quickly in smaller quantities, like spinach or broccoli, and only as needed.
  - Don't stock up on perishable foods.
- Buy things like potatoes and tomatoes individually instead of in a bag if you may not eat them all.
  - Reusable produce bags will also eliminate the packaging.
- Consume food in the order you bought it. Eat the food that is due to go bad soon.
- Don't assume that food past its expiration date is no good. These dates are not regulated, and many foods can still be eaten after their *Use By* date.
  - Look it up and know the truth.
  - If it doesn't look spoiled, it's likely okay to eat.
- Fruits and veggies that are going soft are great for making smoothies.
- Find out the best way to store produce. Some last longer stored at room temperature, like onions and tomatoes.
- Don't overfill your fridge. Too much food means you can't see what's in the back.
  - It's out of sight, out of mind until it stinks.

If you ate out, don't forget about the leftovers you brought home in a reusable container.

- Leftovers make great lunches and save you money.
- Use a marker to label the container with what's in it and the date.
- Put a reminder on your phone if you often forget to eat your leftovers.

Common sense says if we waste less, we buy less, and we reduce food production waste and costs.

One study showed that restaurants participating in food waste reduction programs recognized a $7 savings in operating costs for every $1 of food not wasted.

Less stuff is good for the planet and the economy

Have you heard the expression *less is more*? This is important to practice during our climate crisis. The less we buy, the more life we will get out of our planet.

- The mentality that more is good is dangerous to life on this planet.
- Excess consumption doesn't support a prosperous economy. It only benefits one percent of the population and hugely contributes to our climate emergency.

It's time to resist consumerism. It is a way of life we can no longer afford.

Tip 39 - Rent or buy used equipment and tools to reduce emissions and your footprint.

The manufacturing of new products is responsible for one-third of carbon emissions. This doesn't include the damage done by the waste it creates. One way to decrease this impact is to stop buying new items.

- Tools and equipment you need can occasionally be rented.

- - Like a rototiller to start your garden, a power washer to do your windows or a sander for a furniture project.
- You can rent things like party decorations, formal wear, audiovisual equipment, dishes, and sporting goods, to name a few.

Whatever it is, you can likely rent it. As a result, you will save money, reduce emissions, and not have to store another item you won't use very often.

Save money and the planet by shopping thrift stores, garage sales, or resale websites. With a bit of effort, you can find what you need, spend a lot less on it, and cut out the emissions and waste caused by making new products.

Tip 40 - Clean out your storage to save money, reduce waste, and help others.

You may not think that getting rid of things you have stored can help the planet, but it can. Having items in storage wastes energy and is a source of emissions we have control over.

- There is enough self-storage in the US alone to provide each person with six feet of living space.

Americans spend $38 billion annually on storage lockers. The cost is even higher when we consider the environmental impact of the energy used by these facilities.

- Make it your next project to get rid of items you don't use.
- Go through boxes or your storage locker. Fill your vehicle and drop things off at your local Goodwill or thrift store.
  - I like donating to charitable organizations like Goodwill, where the money goes to help people get back on their feet.
  - You get a tax write-off and feel good knowing you made those items available for someone who could use them.
- Keep working on it. Schedule a day once a week or once a month to purge your stored items.

- You will reduce your expenses and carbon footprint.
  - Once your storage locker is less full, you can get a smaller unit or eliminate the expense altogether.

Tip 41 - Don't support the making of new clothes; buy used and donate unwanted ones.

Since 2000, the amount of apparel produced globally has doubled and continues to grow.

- The garment industry makes around 100 billion pieces of clothing annually.
  - Over 80% of discarded clothing ends up in landfills or incinerated.

The impact of clothing on our environment is serious.

- Apparel manufacturing contributes 10% of global emissions.
- The water used annually to make new garments is enough to quench five million people's thirst.

I began shopping at thrift stores when I was young and broke, but I stuck with buying used when I started making more money. It's fulfilling to find a barely used jacket or toaster and not pay the environmental price attached to making new.

- You can find a lot of great deals when you shop for used clothing.
  - I've bought many items with original price tags still on them.

We eliminate packaging completely when we buy used goods, and we reduce carbon emissions and waste caused by the making and shipping new products.

- If you need clothes, why not first check the racks at a thrift store or visit yard sales?
- If retail therapy is your thing, this can save you a lot of money.

- There are apps and online groups on social media that are all about buying nothing.
- People with items like baby clothes use these apps to sell or give items to their neighbors.
  - Donate your gently used clothes or sell brand-name items at a consignment shop in your neighborhood or online.

There are responsible apparel companies that offer clothing recycling. Some recycle only their products, but others will take any garments in any condition. They recycle the material into new clothes if an item can't be reconditioned for resale.

Many of these retailers offer collection bins in their stores.

Tip 42 - Make your own natural cleaning and personal hygiene products.

Making your own natural products is a great way to reduce waste and your environmental impact. There are many online resources to help you learn how.

- You can find several DIY recipes for making natural cleaners, like baking soda, lemon juice, or vinegar.

If you make your own bathroom products, you will significantly decrease your waste and live healthier. Many items in your bathroom come in plastic containers and are made with chemicals your body can do without.

- Go online and find ways to make things like hand soap, face and body wash, hand sanitizer, lip balm, and even toothpaste.
- Do a little research and find ideas you'd like to try.
- Schedule a time and make it a family project.

It shouldn't take long, and you'll feel better about your waste reduction efforts when it's become part of your everyday life.

Your carbon footprint - knowing what it is and how to reduce it

It's easy to understand that our carbon footprint reflects how daily activities contribute to greenhouse gases. What's not easy is reducing these activities.

I use the term *carbon footprint* because it's familiar to most people, but I hate it because it was invented by the fossil fuel industry as yet another ploy to distract us from the damage their products do to the earth.

Though it feels like we can't control big corporations, many things in our control are old habits we've developed. We've got to break these habits.

Tip 43 - Buy local to help reduce emissions and boost your economy.

When you buy from large chains, the products are often transported to your area from far away.

- Locally owned retailers are more likely to have locally sourced and manufactured products.
- When you spend money locally, it helps your community thrive. It creates jobs that lower crime, and it reduces your environmental impact.
- Though small retailers might charge a little more, in the long run, the value of keeping money in your community is three to four times what you spend.

Tip 44 - How you drive your car affects the size of your carbon footprint.

According to the US Department of Energy, the idling of personal vehicles accounts for 30 million tons of carbon dioxide emissions annually and wastes three billion gallons of fuel. That's a lot of wasted energy that could be avoided.

- If you are stopped, other than in traffic, don't idle your car for more than 60 seconds. By then, you've already used more gas than it takes to restart it.
- Stop-start technology reduces emissions caused by idling.
  - It shuts down the engine when you're stopped, like at a light, then restarts it when you take your foot off the brake.
  - If your car has this feature, don't turn it off; it reduces your emissions.
- Newer car engines do not need to be warmed up.
  - If you have a fuel-injection engine, it works best if you don't warm it up.
  - The only exception would be in really cold weather when you need to clear your windows and not shiver your way down the road.
- Speeding and quick acceleration wastes gas and increases emissions. Gas engines emit the most carbon during acceleration and produce the least amount when you travel at a steady 35 – 40 mph.
  - Always accelerate slowly, then maintain a steady speed to achieve minimum emissions, and don't speed.
  - When on a road trip, use the cruise control to save gas and emissions.
- Carpool whenever possible. This is a great way to save money and reduce greenhouse gases (GHG).
  - You may build some great friendships.

The high cost of wasting water

With our emissions creating a blanket trapping in heat and the subsequent warming temperatures, our planet is getting dryer, and freshwater is becoming scarce.

Now more than ever, it's important to conserve water.

Drought and water shortages are shutting down farms around the globe. We need to conserve water for many reasons, the biggest being the looming food shortages.

Tip 45 - Simple steps to conserve water, our most precious resource.

You may not think a few gallons of water wasted in your home each day can make a difference to those without food and water, but it does.

It's easy to reduce water waste once we're conscious of where it happens.

- Be mindful of how you turn on a faucet. A low stream will often meet your needs and saves a lot of water
- While scrubbing your hands, turn off the faucet until you're ready to rinse.
- Don't leave the faucet running while brushing your teeth or shaving.
- Shower less. Instead of every day, change to every other day.
- Keep your showers to five minutes or less.
- Wash your hair less. Daily use of shampoos and conditioners is not good for your hair or scalp. Stick to two or three times a week.
- Flush less. Each flush in a standard toilet uses 5 -7 gallons of water.
  - There's no reason to flush every time you pee.
  - Remember this saying:
    If it's yellow, let it mellow; if it's brown, flush it down.
- Remodeling your bathroom? Install a low-flow toilet; they use less than 2 gallons per flush.
- Consider installing a bidet. You don't have to remodel. Portable ones are affordable and attach to a regular toilet. They start around $100. You'll feel fresher and cleaner while reducing your water and paper use.
- Reduce how often you water your lawn. It may look greener the more you do it, but we need that water to grow food.

- Wash your car less, and when you do, look for a car wash that uses recycled water.
  - Wash your own car, so you control how much water is used.

Tip 46 - How to cut down water use when cleaning and washing.

We can often use a lot of water when cleaning the kitchen or bathroom.

- Don't leave the faucet running while cleaning a sink or tub.
  - Wet the surface, turn off the faucet, then scrub. When ready to rinse, use a cup and turn off the faucet between fillings.

When you run washing machines, use conservation settings.

- Some clothes washers have a tap water temperature setting, so you don't use energy to heat or cool the water when doing laundry.
- Never wash your clothes at a higher water level than needed.
- Always run your dishwasher on the water- and energy-saving settings.
  - Skip the pre-wash cycle and use air dry.

Consider washing your clothes less. Clothes don't need to be washed after every wear.

- Every wash breaks down the material, reducing the life of the garment and sending small particles of material, like microplastics, into our water system.
- A good rule is if it doesn't smell bad or look dirty, you can wear it again before washing it.

There's Only So Much Planet to Go Around

Reducing waste is one of the most impactful things we can do for our planet. Humans have been filling the earth with trash for too long. We can fix this by reducing how much there is and creating a waste stream that is no longer a one-way street into landfills, waterways, and air.

Where all this trash has gotten us

Emissions from our trash either enters the atmosphere slowly as it sits in landfills or pollutes the air immediately when it's burned.

- A large chunk of our waste consists of plastics. It's in textiles like polyester and nylon and comes in single-use packaging.
  - It is estimated that emissions from the production and disposal of plastics are 50 times that of a coal-fired power plant.

Trash that doesn't end up in our oceans or get incinerated builds up in our landfills.

- Humans dump over two billion tons of trash in landfills annually.
  - That's enough to wrap a line of garbage trucks around the earth 24 times.

What happens if our garbage keeps growing

With the dangerous amounts of greenhouse gases (GHG) emitted from our garbage and other sources, the damage to our atmosphere is already causing increased severe weather, flooding, fires, poor air quality, and drought.

- This depletes our food supply, increases our cost of living, and puts our health at risk.

- It will continue to get worse until we change what caused it. Us.

Waste is more than garbage; it's all the resources we waste on a regular basis, like food, energy, and water. So we need to get all waste under control and eliminate it.

Discover a Healthier Way to Live When You Reduce Waste

It's not just trash that causes damage to our environment. The extraction of fossil fuels and the resources used to make products are big troublemakers.

Waste-not want-not needs to be our motto.

Changing our habits isn't an option anymore. With the reality of the damage staring us in the face, the only option is to reduce waste. If we stop wasting resources, we will have a better chance of reversing climate change.

<u>Consumerism and waste go hand-in-hand</u>

One of the biggest contributors to waste is consumerism. Like many people in the US, I've spent way too much money buying stuff I didn't need. It took me a long time to break this habit. *Can you say retail therapy?*

- People in the US spend almost twice as much on products as in other countries.
  - One study found that as much as 99% of items purchased are discarded within six months.
- In the US, the average person's carbon footprint is four times that of someone in any other country.

Tip 47 - Cut back on buying new products to reduce the waste of resources.

It's illogical for humans to waste resources we desperately need, especially water, to buy goods we put in storage or discard. Yet we often do this.

We buy items because they're on sale, and we feel we got a deal. But there could be a higher price we pay from the making of it or when we dispose of it.

We're starting to feel the true price of this as farmlands shut down due to water shortages.

- New product manufacturing uses 25% of our freshwater supply.
  - o It's one of the leading causes of GHG emissions.

Consumer demand influences how products are made. The same is true when consumers demand sustainable products.

- Companies will make durable, sustainable products if that is all we buy.

We have more to gain than the health of our planet

A disposable lifestyle can be more expensive than we realize. The true costs are spread over time and the life cycle, cradle to grave (or landfill), of the products we buy.

- It's in our increasing health care costs in a contaminated world where poor air quality and toxins continually make us sick.
  - o The World Health Organization estimates that over 7 million deaths a year are caused by poor air quality and deteriorating living conditions.

- We see it in the loss of lives and the damages caused by increased floods, droughts, fires, and food shortages.
  - In 2020, natural disasters carried a price tag to taxpayers of $210 billion worldwide.
    - An increase of more than 26% from 2019.

The waste reduction ideas from this chapter can help us make the adjustments needed to stop and maybe reverse these damages.

1. Pay attention to where you create trash and recycle so you can reduce it.
   a. Pick one disposable item and find a way to eliminate it from your life.
      i. Then pick another and keep going.
2. Cut back on carry-out eating; carry lunches and practice meal planning.
   a. Look for reusable food storage solutions. It's cheaper than disposables.
3. Encourage friends and coworkers to reduce waste.
   a. Make sure they see you doing it.
4. A zero-waste lifestyle is doable.
   a. Start by eliminating packaged and take-out foods, then move on to other sources of packaging that won't recycle.
   b. Always practice zero waste when entertaining.
5. Manufacturing wastes a lot of water and energy.
   a. It's one of the top three sources of greenhouse gases.
   b. Don't buy products you will throw out in a few months.
      i. Buy stuff you want to keep.
   c. Awareness is half the battle to reducing wasted resources.
6. Reduce energy use and water waste to slow down our demand for natural resources.
   a. Turn off faucets more often. Unplug electric devices when not in use.
7. Reduce the amount of food and goods you waste to significantly decrease greenhouse gas emissions from landfills.
   a. Food waste creates significant amounts of methane, one of the most damaging greenhouse gases we emit.

Chapter 3: Reuse it! A Great Way to Save Money and The World

Reuse is the most planet-friendly thing we can do, and most people will do it if given the opportunity. But unfortunately, we don't yet have the systems in place for a reusable lifestyle.

We need to learn how to reuse what we have. We need processes in our supply chains that move us away from single-use by providing reusable options.

Returning to a Life Where Reuse Was What We Did

A few generations ago, people threw out very little.

- Disposables weren't a thing, and food waste went to the animals or was composted for the garden or the crops.
- When things stopped working, they were repaired.
- There was a lot less packaging, and most of it was reusable.

Fast forward to now.

- Every day, enough food is wasted to feed three people.
- Products are thrown out as soon as they don't work or we're bored with them.
- The average person in a developed country expects products not to last.
  - We've adopted a dangerous disposable mentality that most of us wouldn't have chosen if we'd been asked.

We need to go back to reusing valuable materials. We can start by repairing things like electronics, appliances, and clothing.

- We must find ways to repurpose materials.

- It could be clothing, dishes, or aluminum cans. No matter what the item, it likely has more uses when we're done with it.

I'm not suggesting you become a hoarder. No! That would be counterproductive. An uncluttered home leads to an uncluttered mind, which leads to balance and happiness. *We really need a lot more happiness in the world right now.*

We need to reduce what we buy and reuse what we have.

When trash feels inevitable, which it isn't, it helps to have a reuse mindset from the time you buy something to the moment you are done with it and the packaging.

The history of recycling - originally, it meant reusing

Recycling dates back hundreds of years and was originally about reusing materials and fixing existing ones. It was more like a scrap business of today.

Tip 48 - Buy durable products and repair older ones to save money and the planet.

I have a drill that I believe was made in the 1950s. It belonged to my dad and is still one of the best running tools I have. I'm pretty sure my grandchildren will use it. It was made by Black & Decker. There are so few companies that make long-lasting products. I believe this is a brand that does. I've had one of their kitchen mixers for more than twenty years. I don't expect it to ever quit.

"When you buy something cheap and bad, the best you're going to feel about it is when you buy it. When you buy something expensive and good, the worst you're going to feel about it is when you buy it."
Sasha Aickin on financial advice from his grandmother

There are companies that want to do the right thing and make good quality products. It might take a little research to find them, but it's worth your time, the possible higher price tag, and the waste it will reduce.

- When you need to make a major purchase, like a computer, appliance, or TV, find the brand that lasts the longest and has a good repair record.
    - Call a repair place and ask which brands need repairing less and don't cost as much to fix.
        - It's hard to justify repairing something for the same price as a new one.
- Don't throw things out when they no longer work. Look into repairing them or repurposing the materials.

Tip 49 - Repair and repurpose electronics for the ultimate reuse of our resources.

A lot of precious metals and resources go into making electronics. When we repair or recycle them, we avoid mining and production of virgin materials.

- To recycle electronics take them to an e-waste event or collection site so the materials can be repurposed.
    - Look online for an e-waste event in your area.
    - Many electronics stores have recycling programs. They usually take TVs, old phones, batteries, computers, and appliances.
- Repairing electronics can be a challenge, but it's doable.
    - Check with local repair shops before you give up on it.
    - Post it online for free. Maybe someone can repair it or use the parts.

I'll never forget when my first flat-screen TV stopped turning on. When I called customer service, they said there was nothing they could do, and I was lucky to get eight years out of it. I was ticked off. My previous TV had lasted more than twenty years.

Flat-screen TVs are hard to get rid of. They can't easily be recycled and can't go in the trash because of the contamination they cause in landfills. When mine quit, I posted it online for free, hoping someone could fix it or use the parts. The person who picked it up was familiar with the model and said the power boards usually went out after five years. He would pop in a new one and sell it. Great, but it always bugged me that the manufacturer knew the power board wouldn't last. They made it that way.

Give electronics and the materials in them a chance to be reused.

If your work or hobby requires the latest technology, don't think of your old equipment as garbage. Think of it as a resource that can be reused.

- Sell your used electronics online or at a resale shop.
    - Shop used when you don't need to buy new, like an extra monitor or a printer. You may find something that will meet your needs.
- Give it to someone who can use it.
    - Look for neighborhood giveaway programs and apps where people give and get free stuff. NextDoor, Facebook, buy nothing groups, and freecycle.org are a few.
    - Non-profit organizations take donations of working computers, printers, and other electronics.
- If you have a lot to get rid of, have a yard sale. You may meet neighbors you didn't know and make a little money doing it.

No matter what, don't throw out old electronics. The materials used to make them are very toxic to the earth, and we need them to make more electronics.

Tip 50 - Buy used cars and donate old ones to save more than money.

There are good used vehicles out there with plenty of life left in them. So even if you can afford a new car, why not skip the environmental costs and buy a used one?

- It's usually best to buy a used car from a certified dealer. They are required to make sure it is mechanically sound and emissions are minimal.
- Electric and hybrid cars have been around for a while, so there are more used ones available.

Several non-profit organizations take donated cars and fix them up for someone in need. Find one that supports people you want to help, like veterans or low-income families.

- They will pick up your old car.
- You'll get a tax write-off for the book value.
- Some organizations take any type of vehicle, including bikes, motorcycles, and boats.

The not-so-great place we ended up

Initially, out-of-sight, out-of-mind worked, so we embraced recycling because we could get rid of our trash and not feel bad about it. If it had worked right and what we put in our bins had been recycled, all packaging and single-use plastics would now be made from recycled materials. They're not.

People and the companies we buy from need to work together to reduce waste and reuse the materials we have. One way to do this is to support stronger recycling programs. Another way is to find uses for materials already coming into our homes.

Reusing Stuff is Cool and Can be Fun

While a fantastic way to reuse stuff is to buy it used, there are also many ways to reuse disposable items around your home.

Reuse is the ultimate form of recycling.

There are useful and crafty things you can do with almost every item you put in your recycling and trash. Rather than run out to buy something that could end up in the garbage, first try reusing materials you already have.

Add value to your trash - how to start reusing it today

Change the way you think about your trash. Look at it as a resource. Something you can use to make life easier and healthier.

Tip 51 - Save bubble wrap and foam packing peanuts to reuse for shipping and moving.

It's never easy to know what to do with packing peanuts or bubble wrap you get with packages. These can't be recycled and are very toxic to the planet if thrown in the trash. But they can be reused.

- Keep a bag or box of packing peanuts and bubble wrap handy.
    - For when you ship things to friends or family.
    - If you move often, save them for the next time you pack up.
    - Use them to store breakables like holiday ornaments and keepsakes.

If space is limited and you end up with more than you can store, make sure to get rid of it responsibly.

- Never put foam packaging or bubble wrap in your recycling bin or trash.
    - Bubble wrap may be accepted at stores with collection bins for shopping bags and other soft plastics.
- Shipping and mailing stores often take packing peanuts and bubble wrap. They may even take big pieces of foam you get with appliances.
    - Do you know someone who is moving soon? Maybe they could use it.

Tip 52 - Don't buy storage bins; reuse empty shoe boxes and other containers.

It helps to organize things in boxes you can label. Don't run out and buy rubber or plastic storage boxes. Instead, reuse shoe boxes or small containers.

- Shoe boxes are great for organizing and storing documents or nick-nacks. For example, turn one into a sewing box, or store crayons, paints, and craft supplies.
- You may have empty containers like baby wipes with lids that snap closed. These are great for storing things like crayons, markers, crafts, and sewing items.
- Save shipping boxes. Break them down and store them in the back of a closet.
    - When you organize tools or crafts or ship something, you're bound to have the size you need.
    - Make life easier and keep them in order by size; big boxes in the back, small ones in the front.

Beyond handy reuses for items you may otherwise dispose of, there's also a lot of cool stuff you can make.

Tip 53 - Worn out clothes have many uses around the house.

Clothes too worn to donate can be cut up to reuse the material.

- Old T-shirts make great dusting rags or use them to clean mirrors, and windows, wash your car, or wipe up messes.
- Use old nylon or polyester to line your cat litter or as a pet bed liner to replace disposable ones.
- Sew the front and back of a shirt together, then stuff it with packing peanuts to make a dog or cat bed.
- Use worn materials for gardening.
    - Cut material that is worn thin into pieces to cover the drain hole in your planters.
        - This prevents dirt from falling out while still providing drainage.

- o Use a rubber band to attach a piece of cloth to an empty detergent bottle. Poke holes in the cloth to make a watering can.
- If you love crafts, cut up old clothes to decorate walls and planters or make squares for quilting.

Tip 54 - So many ways to reuse glass, metal, and plastic containers.

It's hard to avoid buying disposable containers, but many have great reuses around the home. You will be amazed by how you can reuse your trash, from tools and decorations to crafts and gifts.

- Pickle jars or any wide-mouth glass jar can be used for mixing.
  - o The lids seal tight so you can shake them to mix ingredients.
    - ▪ What a fun thing an older kid can do to help in the kitchen.
  - o Use them to mix and store paints.
  - o A great way to store a number of things, like nails, thumbtacks, marbles, or buttons.
    - ▪ These are handy when you want to see what's in the container.
- Baby food jars and yogurt cups with lids are great for storing small items.
  - o Write what's in it to keep things organized. Use masking tape if needed.
  - o These are great for storing screws and nails or bathroom items like cotton swabs and bobby pins.
  - o Decorate them with paint or repurposed cloth to match your decor.
- Make a snack container out of plastic wide-mouth juice bottles. Wash and dry them to fill them with snacks like nuts, small crackers, granola, or raisins.
  - o These can be dropped in lunches or carried on outdoor adventures.
- Plastics are so durable that you can make bricks with them.
  - o Look online for how to make an eco-brick and use plastic bottles for landscaping.

- What a great way to save money when you landscape your yard and they look cool.
- Plastic tubs with lids that butter or cottage cheese come in make great storage. These are number 5 plastics that aren't easily recycled.
  - Why buy food storage containers? Reuse plastic tubs.
  - Use them to store the plastic bags and wraps you washed for reuse.
  - These tubs are great for painting. The lid usually provides a tight seal so you can store your paint for a short time, and it won't dry up.
  - Tubs are handy for storing craft materials or crayons, and they're easy to label.
- If you haven't eliminated carry-out eating, you may have an occasional plastic or foam clamshell container. Since these can't be recycled, there are ways to reuse them.
  - Carry sandwiches or salads for lunch.
  - Make a plant terrarium out of plastic clamshells, cut off the top, and use both pieces to start seedlings.
  - Store craft supplies, like markers and stickers. It makes life easier when they're in a see-through container.
  - Use them as a paint pallet. The bigger ones work great for this.
- Old detergent bottles can be turned into useful household items.
  - Cut down the middle, and with a rubber band and an old hair brush, you can make a scrubber.
  - Turn them into a watering can for your plants. Attach used aluminum foil or old cloth over the opening with a rubber band and poke holes in it.
  - Cut off the top to make a storage bin. The ones with handles are great for this when cut at an angle.
  - There are so many creative reuse ideas for detergent bottles online. I'm sure you will find something you can do with them.
- Steel cans, like for soup or coffee, can be used to make planters, pen holders, lamps, or even a wine rack.
  - Make a pen holder. Paint the can or glue on paper or material from old clothes. Use different colors to make letters and personalize it.

- o Make your own planter. Drill holes in the bottom of a can and decorate it to your style.
- o Make a personalized wine rack. Cut off both ends of a few cans, decorate them, and glue them together. This a great gift idea that you can personalize.
- o Making your own furnishings. Look online to find ways to make things like lamps, wall art, and even furniture.
- Aluminum cans are handy for crafters and bakers. This soft metal is flexible and can be used to make crafty tools.
  - o With heavy-duty scissors and gloves on, cut the can to remove the bottom and top, leaving you a ring of aluminum you can shape how you like.
    - Sand and tape the edges for safety.
    - Get creative by gluing shapes together or cutting the aluminum differently.
  - o Make your own cookie cutters in personalized patterns.
  - o Design stencil patterns that are fun for decorating a wall or clothing.
- Propane cans can be refilled and reused over and over. Hardware stores will refill most types.
  - o These can't go in recycling or trash and need to be properly disposed of if you don't reuse them.

Tip 55 - Great ways to reuse newspapers, flyers, and cardboard.

These common everyday items can be reused in many ways around your home.

- Clean your windows with newspaper to prevent streaking.
  - o Use a vinegar and water solution for an environmentally friendly way to clean your windows and mirrors.
- Use newspaper to wrap gifts instead of buying wrapping paper. The comic section can make the gift extra fun.
- Packing up for a move? Collect newspapers and flyers to wrap dishes, photos, and other breakables.

- Use cardboard or cereal boxes to make spacers for packing glass and fragile items.
- For the outdoor adventurer who likes to camp, make fire starters with cardboard and string.
  - Cut cardboard into squares about the size of your hand, roll it up and tie a string around it. Leave one end of the string a little longer to light it. Put a few under your woodpile to get the fire started.
    - Melt candle remnants for the wax to dip these in, so they burn longer.
  - This is a fun way kids can help prepare for camping trips while learning about reusing materials.

Make fun things you and your family will enjoy

Studies show that doing something creative is great for your mental health. This is why people say that doing crafts relieves their stress and anxiety.

- Most people report elevated happiness when doing a craft.
- Creativity is crucial to a child's brain development.
- Doing crafts benefits the entire family and provides bonding time with kids.

Tip 56 - Cool things you can create with paper towel and toilet paper tubes.

The cardboard tube at the center of the paper towel and toilet paper rolls can be used to make fun things.

- Decorate them to make holders for power cords and cables.
  - Label them, so you know if they contain phone chargers, power cords, TV cables, etc. Then you know what's in it when you need one of these.
- Make napkin holders out of decorated cardboard cores.
- Decorate and add streamers to turn them into a windsock or connect aluminum and metal strips to make a wind chime.

- Make a castle, village, barnyard, dollhouse, or racetrack for kids to play with.
  - They can decorate these tubes with paint, markers, construction paper, and glitter.
  - Cut and glue them together to make walls, roads, bridges, and tunnels.
    - The cardboard becomes stronger once painted.
  - Add to the design by cutting shapes from construction paper or pictures from magazines.
  - Make animals and people by adding pipe cleaners, paper clips, or cut-up straws.
- Kids love to explore music, and making instruments is easy with cardboard tubes and other items you have around the house.
  - Put rice in a cardboard tube and tape paper over both ends. Decorate it, and you will have a rain stick.
  - Put wax paper over one end of a tube with a rubber band. Decorate it and hum into the open end. You've made a homemade kazoo.
  - Cut a hole in the center, wrap a few rubber bands around it, and you have a mini guitar.
    - Use an empty tissue box to make a bigger guitar.
  - There are a lot of cool instruments you can make with materials you have around the house or in your recycling bin. Look online for ideas.
  - Your kids will love coloring these and designing their own instruments.

The most fulfilling crafts are the ones that don't cost money and teach kids how to reuse waste. Challenge kids to come up with creative ways to reuse items in the recycle bin. They may surprise you with some great ideas.

Tip 57 - Make your own holiday decorations and other fun and useful crafts.

Don't run out to buy decorations for your home until you explore eco-friendly ideas to make your own. You'll enjoy your decorations more if they're your own unique design or your child made them. In addition, your friends and family will love getting handmade gifts from your kids.

- When a board game is worn or missing pieces, use what's left for making decorations, jewelry, crafts, and games.
    - o Drill holes or get a hot glue gun to turn game pieces into jewelry or ornaments.
    - o Repurpose game pieces and playing cards.
        - Make furniture or characters for a playhouse or barnyard.
        - Make jewelry for kids to play dress up or put on their dolls.
        - Build a house of cards. It's fun to see how big you can make it before it falls.
            - What a fun activity for the whole family.
    - o Decorate playing cards, then run a thread through the top with a needle to hang them. Add some glitter, so they sparkle.
        - Cut them into creative shapes or objects.
        - Decorated by your kids, these make beautiful ornaments and gifts.
    - o Add glitter and hang game pieces from anywhere light will make them sparkle.
        - A beautiful addition to any yard.
    - o Cut up an old game board to turn it into something else.
        - This coated cardboard can't be recycled but may be repurposed to make decorations or useful things, like coasters.
    - o Save dice to use with other games or for your favorite dice games.
- Cut aluminum cans into strips to make ornaments.
    - o Sand down sharp edges before decorating.
    - o Spray paint and glitter are a great way to decorate these.
        - Use a piece of wire or a hot glue gun to gather them into a sparkly bunch.
        - Twist them into spirals before you hang them.

There are a lot of cool things you can make with items you might otherwise throw out.

- Make floating drink coasters out of wine corks for pool time.
- Create wall art, picture frames, herb gardens, desk organizers, napkin rings... there are ways to make these and a lot more.

When you make your own decorations or your kids make them, you're less likely to throw them out.

Tip 58 - Crafty ways to reuse a collection of plastic shopping bags.

There are so many creative things you can make with an accumulation of plastic bags. I was blown away by all the cool things I found online.

- Make a holiday wreath. Some designs are easy to make and are prettier than one from a store.
- Make garland for your holiday tree. Knot the ends together; cut them into strips and braid them first for a different look. Then, lightly spray with glitter paint to add sparkle to your tree.
- Make a kite by cutting a plastic bag into a triangle, adding two sticks and some string. The instructions are online.
- You can tie or braid shopping bags together to make a rope.
  o Add handles or simply knot the ends, and you've made a jump rope.

Plastic is durable and lasts a long time. That's why we have so much of it. When soft plastics are layered together, they become stronger. This is why a lot of creative things are being made with them.

Tip 59 - Egg cartons are great for crafts and other cool things, like planting and games.

Egg cartons have lots of potential reuses around the home.

- Store small items in them like beads, jewelry, picture hooks, and sewing supplies.
  o Decorate them with markers or paint to fit your motif.

- A fun craft for kids is to cut out the cups and glue two together to make an egg shape they can decorate.
  - Make holiday decorations. They're easy to decorate and lightweight, so they can hang on a tree.
  - Paint faces on them to make little people.
  - Cut the flat top into shapes kids can decorate to make tiny furniture.
  - Make replacement pieces for your board games.
- When you or your kids paint, use egg cartons as a palette to keep colors separate.
- Cut them up to use as a paint tool to make interesting shapes on your canvas.
- Use them for plaster molds. Drop beads in the cups, then pour in the plaster.
  - Kids can put their initials on the bottom.
  - Once dry, pop them out of the mold and paint them.
- Use them to start seedlings or make a bird feeder for your garden.
- Use them as shipping or packing material to prevent breakage. They're great for storing glass ornaments or other small breakables.
- Make a fun travel game for road trips and camping with an egg carton and marbles or pebbles.
  - Look up Egg Carton Mancala for instructions on how to make and play this ancient two-person strategy game.

Tip 60 - Reuse materials to make beautiful jewelry for you and your kids.

There are so many small items you may often throw away that can become something beautiful. Here are some things you can use to make jewelry for you and your kids.

- Extra buttons can be strung together to make a necklace or bracelet.
- Plastic lids, bottle caps, and wine corks can make unique earrings or necklaces.
- Paper clips can be formed into any shape to hang from your ears or link together to make a necklace.

o Wrap the shape in leftover fabric or ribbon. Then, create your own unique designs.

Always look for crafts you can do with the materials you have before buying new supplies.

Reuse to Save the Planet and Create a Better Life

Many activities and hobbies can incorporate reusing materials in your home. This is a great way to save money and create a better life for you and your loved ones.

When you do this, you get that - so many uses for wasted materials

As you start to pay attention to what goes in your trash and recycling bin, look for ways to get more value out of it.

Tip 61 - Disposable items can be reused to beautify your yard or garden.

From beautiful designs to handy tools, there are many uses for disposables when you're gardening or landscaping.

- Turn large plastic bottles into planters. Cut off the top of detergent bottles, milk jugs, or two-liter bottles. Decorate and drill holes in the bottom for drainage.
  - o Make sure to sand the edges where you cut the plastic.
  - o Fun for your kids to decorate and start gardening with you.
- Cardboard and newspaper make a great weed barrier.
  - o Put wet newspaper down before laying a rock bed, bricks, or woodchips, and you'll have fewer weeds.

- Turn a detergent bottle into a watering can. Secure cloth or foil over the top with a rubber band or twist ties and poke holes in it.
- Plastic bottles can be turned into a stool or plant stand for your patio. Patterns are available online.
  - Plastic is durable and makes great outdoor furnishings that hold up in any weather.
- Aluminum is great for making yard art since it bends easily and can be shaped into all sorts of things, like a windmill or a bird. It sparkles in the sun.
  - When you cut up aluminum cans, save the bottom for planting seedlings.
    - Collect a few of these and glue them together to create a succulent garden.
- Use colored glass to make a mosaic and decorate your home or yard to add color and shine.
  - Put a few different colors of glass in an old pillowcase to lightly smash them into big pieces. Very therapeutic, but use caution and wear safety glasses.
    - Stick big colorful pieces into the ground to decorate flower beds.
    - Fill a clear vase or plastic bottle and put it in your flowerbed or garden to add sparkle.
    - Add color to a clear vase before putting flowers or a plant in it.

Tip 62 - Plastic containers lighten the load and add drainage when canister gardening.

I've done a lot of canister gardening and was excited to learn this trick. Dirt's heavy!

- Before you put dirt in the canister, add empty plastic containers or soda bottles.
  - With these in the bottom, your canisters are lighter and easier to move. And it improves drainage.

Tip 63 - Compost your food waste to grow healthy plants.

The best way to handle food waste in your home is to compost it. This organic material contains nutrients your soil needs, like nitrogen, phosphorus, and potassium, that plants thrive on.

- All organic matter contains carbon dioxide (CO_2). Composting returns this greenhouse gas to the soil where we need it.
    - You shouldn't till or turn your soil. Instead, leave the nutrients and the CO_2 in the ground.

Nowadays, composting is easier than it sounds.

- There are portable composting units that fit on your kitchen counter. These are great if you don't have the outdoor space for a big composting unit or live in an apartment.
    - Portable composters are small enough to fit on your kitchen counter and will turn your scraps into plant food overnight.
- If you have yard space, outdoor composters are easy to set up and use.
- Build your own composter. You can find instructions online and likely have some of the materials around the house that you can repurpose.

Don't put things in your compost that will harm your plants. Chemicals or toxic materials, like plastics, don't belong with your food waste.

- Paper is not always compostable because it can contain inks or dyes.
    - Only compost plain white paper towels and napkins. Color or printed ones go in the trash.
    - Printer paper, newspaper, junk mail, etc.... should be recycled, not composted.
- If you have a portable in-home composter, follow the instructions as to what you can put in it.

Tip 64 - Reuse food scraps around the home to save money and make life easier.

There are a lot of handy reuses for your food scraps.

- Put used coffee grounds in a plastic tub you've rescued from recycling and set it in the back of your fridge to absorb odors. Why waste baking soda?
  - Replace it every few weeks and use the old grounds as an abrasive by adding it to dish soap when tackling burnt and stuck-on foods.
- Used tea bags can be beneficial to your health. Store them in the fridge in a repurposed container with the lid on. Use them for puffiness and inflammation.
  - Cooled tea bags placed on your eyes in the morning can reduce puffiness and help wake you up.
  - Put them on insect bites to reduce itching and swelling.
- If you're a juicer, sprinkle a little salt on a piece of juiced citrus to clean silverware and tarnished surfaces.
  - The acid in the citrus combined with salt makes a great natural cleaner for hard-to-polish items.
- Kitchen grease can be hard to dispose of since it can't go down the drain.
  - Reuse a wide-mouth jar, like a pickle jar, for your bacon grease or meat fat, then reuse it for cooking. This is how they cooked a few generations ago before processed cooking oils came along.
  - Make a bird feeder by putting used grease into a shallow can, like for cat food, and freeze it. Then hang it in your yard for birds to eat. Certain birds love cooking grease.
    - Drill holes before putting the grease in, or glue some string to the sides for hanging it in your yard.

The scrap business - reusing waste has been around for a while

The scrap industry has been around for a long time and contributes over $100 billion to the US economy each year, $4 billion of which goes into local economies.

Tip 65 - There may be cash in your unwanted scrap metal and appliances.

Metal is one of the hottest commodities on the scrap and recovered materials market, so scrap yards will often pay for it.

- Search on how to get paid for scrap metal in your area. There are resources that help you get the most money for it.
- A great way to make money from old appliances is to repair and sell them. Or you may be able to sell the metal or parts.
  - Not much of a fix-it person? Call scrap yards in your area. They'll pick it up for free or may pay you for it.

Tip 66 - Find a scrap exchange, shop resale, and donate good condition items.

I grew up in a town where we had several resale and consignment shops, and we had a scrap materials exchange. People would bring in items they had an excess of, like baby food jars, cloth, pipe cleaners, buttons, CDs (which make beautiful ornaments when decorated), and even stickers. They always had stickers.

As a Scout leader and Sunday School teacher, whenever I had a craft in mind, I would go to the scrap exchange and find most of the materials I needed for cheap. They kept their doors open by selling donated materials. Check and see if you have one in your community.

- In California's Bay Area, the town of El Cerrito offers residents an Exchange Zone where people donate everyday items, and their neighbors take what they need.

- o They also have a reuse station where things can be donated for sale through a thrift store that supports their non-profit operations.
- One of the friendliest ways to reuse things is to buy used products locally. I've met many neighbors going to yard sales or other teachers and leaders at the scrap store.
  - o People often post online their rummage sales or items they are selling.
  - o There are apps to post items for sale to people in your neighborhood.
- Buying used and sharing goods within your community has an economic impact.
  - o If you donate items to a thrift store, they're sold locally. This supports jobs, keeps the money in your community, and eliminates shipping emissions.
- Resale shopping is a great way to save money and the planet. Maybe earn some cash, too.
  - o If you have valuable brand-named items, try selling them at a consignment shop.
  - o Online consignment shops are available where you can sell brand-named apparel.

The scrap business is expected to grow significantly over the next decade as our awareness of the climate crisis increases, and we work to reuse materials.

Buy Recycled - The Best Way to Reuse What We've Got

Buying products made from recovered materials, like plastics, is important to fighting global warming. It will clean up our garbage and improve the value of recycled materials as demand increases.

Tip 67 - Buy products made with post-consumer content to reuse existing materials.

Before you purchase a needed item, check to see if you can buy it with recycled content and packaging. When you choose to buy these products, it accomplishes two things.

- It reduces the trash going into our oceans and landfills, and
- Supports better recycling, so more recovered materials are available for purchase.
    o The more we buy recycled packaging, the more it will become available. Prices will drop, and more manufacturers will use them.

The best support we can give our planet is to buy from companies making products using post-consumer recycled (PCR) materials. This is how we get circularity with our waste.

- Technology companies are making phones, laptops, and other hardware with PCR material.
    o Check your favorite tech company's website to see if they do.
- Many apparel companies offer clothes made with recycled materials.
    o Polyester, nylon, spandex, and rubber are made from fossil fuels. Recycled plastics need to be used to make more of these, not virgin.
- Parley for the Oceans is a non-profit focused on cleaning up and protecting our waters. They partnered with a leading sneaker company to offer a running shoe partially made from upcycled ocean plastics.

We have the materials and the technology to make our waste circular. But we need the demand to drive it.

The ultimate reuse - recycled packaging promotes circularity

Circularity is important to eliminating single-use packaging. If we can recover and reuse materials that have already been extracted from the planet, we...

- Reduce emissions from the mining and manufacturing of new packaging.
    - Yes, there are emissions that go with recycling, but when you factor in hauling and shipping garbage and the damage it causes to our planet, recycling has less of an impact.
- Stop depleting our natural resources that have been over-extracted.
    - Drilling, deforestation, industrialized farming, and livestock are causing irreversible damage to our planet.
        - It doesn't make sense to keep doing these things when alternatives are available.
- Eliminate our one-way waste system that fills our planet with trash.

Tip 68 - Demand recycled packaging and use the power of your wallet for change.

Due to a lack of demand, the recycled plastics market has been slow to grow. This means these materials are more expensive. As a result, manufacturers will only switch to them if they are forced to by consumers and government regulations.

Government incentives for companies that use recycled packaging would help, but some governments are still subsidizing the fossil fuel industry that profits from making virgin plastics. This helps keep virgin plastics cheap and recycled plastics off our shelves.

We can't wait for laws requiring companies to use recycled packaging; it's on us to demand it. Customer demand is the quickest way to get companies to change.

- Look up brands using recycled packaging and start buying them.
- Visit websites for your favorite brands and write emails telling them you like their product but that you would prefer they use recycled packaging.
  - If they receive enough of these requests, they will realize they need to do things differently.

We've already generated enough trash. I hope you'll use these points from this chapter to reduce waste and slow down demand for new materials.

1. Buy products that last. Repair or repurpose old electronics and equipment.
   a. Always dispose of electronics at e-waste events so rare metals and other components can be reused.
   b. Find a scrapyard for metals like tools and appliances when they no longer work.
      i. Reusing these materials will reduce emissions and waste from mining new.
2. Buy used vehicles and equipment to avoid emissions from making new.
   a. Most equipment and tools can be rented.
      i. Some libraries now have loaner programs for these.
3. Find ways to reuse materials before you toss them in the garbage or recycling.
   a. Spend quality time with your family while you make crafts and tools with disposable items. You'll save money and reduce your impact on Earth.
4. Buy used goods and donate good condition items to thrift stores, give them away, or sell them through resale channels.
   a. There is always someone out there who can use your unwanted items.
   b. No matter which you choose, there's an app for that.
5. Buy products and packaging made with post-consumer recycled (PCR) content.
   a. This is how we will build a circular waste system.
   b. Our current one-way system of trash must stop.

Chapter 4: Recycling - Back to Basics, Let's Get it Right

Recycling sucks! There, I said it. You've probably said the same thing.

It's not the act of doing it that sucks; that's important and feels right. It sucks because it's confusing, and it doesn't seem to make a difference. I read recently that two-thirds of Americans surveyed said they were confused about how to recycle.

I get it. I've been preaching recycling for decades, only to find that not everything I said was true. It often changes, and every community does it differently. Plus, the instructions we get from our recycling service can be confusing.

Tip 69 - Stay current on what can go in your bin and what will be recycled.

Recycling rules change as we work to fix the problem. For example, in California, new laws will limit the use of the recycling triangle to containers that actually can be recycled. This will change what's allowed in bins.

While laws and regulations can help fix recycling, they don't keep non-recyclables out of your bin. Only you can do that.

Regularly check the website of your recycling service. Put it on your calendar to do it once a month.

- If you know the name of your service, great. If not, check your local government website.
- Some communities have pictures or lists of acceptable items on the bins.
  - These may not always get updated, so check online to verify what is currently accepted.

Ultimately, we want to reduce and reuse. But when we do recycle, it's important we do it right.

"But I Recycle!" How to Do It Right So It Matters

Humans aren't stupid. We've always wanted recycling to work and understand why it's important. No one believes we have infinite space to store our trash. Studies show most people want to do their part and recycle. The tough part is figuring out how.

How to recycle is pretty easy:

    Step 1 - know what disposables go in your bin

    Step 2 - make sure they're clean and dry

    Step 3 - put them in recycling at your home, office, or drop-off station.

For the most part, we try and do this. The problems come when we don't know what should go in our bins and what happens to those materials.

Most recycling services are only a Materials Recovery Facility (MRF) that sorts and sells what we give them. They respond to market conditions as to what they can or cannot sell, and that often changes.

We need incentives and legislation to build systems for disposables to get recycled, and we need to push governments to do this. In the meantime, recycling correctly is important, so materials we put in our bins have a chance of being reused.

Tip 70 - Don't crush or flatten bottles, containers, or aluminum cans, only cardboard.

I used to think it was helpful to flatten my recycling to save space. Now I know this probably prevented those cans and bottles from being recycled.

- Key to single-stream recycling, which is what most of us have, is that we let someone else sort the materials. Or something like a robot.
  - Sorting is done by sensors that detect an item based on its size, shape, and what it's made of.
  - If crushed or flattened, sensors can't tell what it is, and it ends up as trash.
  - Cardboard and paperboard, like cereal boxes, are the only things you should flatten before putting them in your recycling bin.
    - Check with your MRF; they may want cardboard bundled.

What's easy to recycle and how to do it correctly

Where I worked back in the 1990s, I would go around and put a box under everyone's desk and ask them to put used paper in it so I could take it to recycle. One day someone asked me, "What difference will it make if I recycle this one piece of paper?"

It's not that people don't care, but they often can't see beyond a single piece of paper. Think about it; if it was a drop of water, it would make a difference if you were dying of thirst. So while it might take ten or twenty drops to wet your whistle, wouldn't every single drop matter to you?

Every single bit of waste prevented matters to the planet.

Tip 71 - Always recycle everyday paper like printouts, newspapers, and unwanted mail.

In the US, about 70% of office paper is recycled. Paper recycling is a habit we've been building for years, and it's effective because there's a market for recycled paper, and it's easy to identify in the materials recovery process.

- Paper clips and staples are easily extracted and won't prevent paper from being recycled.

- You might want to keep paperclips and clamps to reduce the need to buy more.
- Shredded paper can't go in your bin. It's too fine and won't get recycled. See **Tip 77** to learn what you can do with it.

For paper to come back around and get recycled, it needs to not get wet or contaminated by food. That's why other items in your bin must be washed and dry.

We need to keep putting paper in our bins and buy it with recycled content. This saves trees that naturally sequester CO_2.

Tip 72 - Cardboard is everywhere, and most of it's easy to recycle.

Most cardboard is recyclable, but some of it is mixed with other materials and may not be allowed in your bin.

- The most common cardboard to get recycled is shipping boxes.
  - Always break these down to flatten them, and follow instructions from your recycling service.
  - Your service may ask you to remove tape and labels. It is always good to do this since it improves the chances of it being recycled.
- File folders and wrappers from paper reams that aren't plastic go with mixed paper recycling.
- The cardboard cores from your toilet paper and paper towel belong in recycling, or see **Tip 56** for ways to reuse them.
- Those handy cardboard cup carriers you get with a big beverage order can go in your bin or back in your car to reuse for the next trip to the coffee shop.
- Containers from dry goods, like cereal boxes, cracker boxes, and cardboard egg cartons, are okay in your bin.
  - The freshness bag inside your dry goods box can't be recycled. It's made of wax or soft plastic and belongs in the trash.
  - Only cardboard egg cartons can be recycled. Don't put foam cartons in your bin. Polystyrene foam can't be recycled so avoid buying it.

- Frozen food containers are coated with plastic and won't recycle.
  - Another reason why eating frozen foods is bad for us.

Tip 73 - All metals are recyclable, but not all can go in your bin.

Metal recycling rates vary by type. For example, about 70% of aluminum is recycled, while steel and copper have a lower rate of around 50%. We need to get better at this since recycling metal is more important now.

We need metals for clean technology like batteries and wind turbines, and recycling metal saves energy. It uses 70 - 95% less energy to recycle metal rather than make new.

- Wash all single-use metals, like soup, coffee, and soda cans, and put them in your recycling.
  - If you have bottle deposit laws in your area, please participate in these since they improve the chances a can or bottle will get recycled. See **Tip 96**.
- Aluminum foil that is clean and dry can go in your recycle bin.
  - If there is oil, grease, or food residue on it, put it in the trash.
- Aerosol and paint cans usually need to be taken to a hazardous waste drop-off.
  - If your MRF says they take these, make sure the container is completely empty and recycle it according to the instructions provided.
    - Aerosol cans can explode and injure workers if not empty.
  - Some residential services take empty paint cans. If yours does, remove the lid and make sure the paint is completely dry before putting it in your bin.
    - If you have leftover paint, take it to a hazardous waste drop-off, or your local paint store may take it.
  - Paint and aerosol cans never go in the trash; they can be toxic.

- If it's not disposable, it can still be recycled, but not in your residential pick-up.
  - Things like fire extinguishers and propane tanks can be recycled but require special handling, even if empty. Take them to your local hazardous waste site.
    - Propane tanks are usually made to be refilled. Check with a local hardware store. If they can't refill the type you have, they may know where you can recycle it.
  - Household items like pans, tea kettles, silverware, and ironing boards have metal that can be recycled, but not in your residential bin.
    - Look for a scrap metal business in your area that will take these.
    - If an item is still working, don't throw it out, donate or sell it.
    - Many communities have special pick-up days for large items and things that aren't accepted in your recycle bin. This should be your last resort since these items may end up in a landfill.
- Copper is in huge demand.
  - There's likely a scrapyard in your area that will pay you for it.

Unlike plastics and paper, metal does not degrade and can be recycled over and over. Moreover, it has an infinite lifecycle making the circularity of metals easy.

Tip 74 - Glass is recyclable, but only the single-use kind like bottles.

Single-use glass can be recycled an infinite number of times. But the amount that gets recycled is low. Several things can cause the single-use glass not to get recycled, even if you put it in your bin. Food contamination and the color of the glass are big factors.

- Glass that isn't clear - minerals were added to color it - can only be recycled with the same color glass. So while your MRF may accept all single-use glass, only that which makes it through sorting and ends up with like colors will get recycled.

- o Buy clear glass whenever possible. It has a better chance of getting recycled.
- Some glass containers are hard to clean, like olive oil bottles with plastic pour caps that are difficult to remove. If you can't clean it out, throw it out.

Non-disposable glass can't be recycled.

- Glassware, like dishes and bakeware, is not single-use, so it can't be recycled. They're treated glass that's not made to be disposable. Same thing with ceramics and any other breakables you may have.
  - o Donate them, sell them at a yard sale, or give them to friends. Don't throw them out unless they're broken.
- Auto glass and plexiglass are not recyclable. Look online for a place where you can take these for proper disposal.
  - o There are scrap services that may find reuses for the material.

Tip 75 - Labels on or labels off; why it's so hard to get it right.

Your recycling service may specify they want labels removed, but most don't. However, removing the label will give a can or bottle a better chance of being recycled.

Even if your MRF accepts containers with labels on them, the people who buy the materials to recycle don't like labels. They reduce how much gets recycled and cause problems with machinery. There's also an issue with the type of label used.

- Soft plastic shrink-wrap labels cause big problems for recyclers.
  - o When you buy single-use beverages, get the ones without a plastic label. This soft plastic can wreck a recycling system.
    - ▪ If you have a container with a soft plastic label on it, cut it off and trash it.

- Glued-on paper labels come off in the recycling process but can cause problems in the machinery.
    - To support better recycling, remove all labels and put recovered paper labels, like from soup cans, in your bin with paper recycling.
    - For glued-on labels, soak the jar in water and remove as much as you can.

Stuff that's not so easy to recycle and what to do with it

Some disposables can be recycled, but not in your bin. These may need to be taken to a drop-off facility or recycled through a mail-in program. There are also items you may think can be recycled, and even if they're allowed in your bin, it won't happen.

I know it's a pain, but here are things that need extra effort from us if the materials are going to get reused again, which is the point of recycling.

Tip 76 - Not all mixed paper and food containers can be recycled.

Not all paper and cardboard food boxes can be recycled since they're mixed with other materials, like plastic and metal.

- Give your window envelopes and tissue boxes a better chance of getting recycled. Remove and throw out the plastic film covering the window or opening.
- Some food containers are more than paper. Milk cartons and other liquid paperboard containers have a plastic coating.
    - There are two types of single-use paperboard containers for liquids.
        - Containers from the refrigerator section are coated with plastic. Many recycling services accept these. Look on their site under cardboard to see if they take milk cartons.

- The cartons on a store shelf that aren't refrigerated have a plastic *and* aluminum coating to preserve the liquid. These containers have too much mixed material to be recycled. Instead, they go in the trash.
  o Frozen food containers are coated paperboard that some recyclers don't take.
    - Plastic trays and wraps that many frozen foods come in aren't recyclable, even if they have a number and a triangle.
  o Take-out food containers made with paperboard are coated and have food contamination, so these can't be recycled.

If your service accepts mixed paperboard cartons, they should specify which ones, like milk or juice cartons, on their website.

- Don't assume they take all paperboard containers.

Tip 77 - Never put shredded paper in your recycling bin.

Shredded paper is considered a mixed paper that requires special handling. Your recycling service may not take it because it gets tangled with other materials and isn't easy to sort out.

- If your service says they take it, they require it is bagged separately from the rest of your paper recyclables. Look for instructions on their website.
- If your service doesn't take it, look for a shred event in your area. They will take what you've already shredded or will shred your documents for you.
- If you have space, save your shredded paper for padding when moving or storing breakables.

Tip 78 - Lids and caps aren't the same, so they don't go in your bin the same way.

Lids and caps are made of different materials. Since they aren't the same, they aren't recycled the same.

- A plastic top goes back on the bottle or container before you recycle it.
  - The best chance a plastic cap has at being recycled is when it's on the bottle. This is because a bottle can be detected in sorting, but small caps can't.
- Metal lids and caps go in your bin separately. These can be tricky to recycle. Use the magnet test to make sure it's metal. If it doesn't stick to a magnet, it can't be recycled.
  - Never put metal lids back on glass bottles. This reduces the chance that the glass will get recycled.
  - Collect metal lids and bottle caps in a can, then recycle them according to the instructions from your service.
  - Metal lids with a rubber or plastic coating inside can usually be recycled with other metal lids.

Tip 79 - How to recycle cooking oil at home and work.

If you pour cooking oil or grease down the drain, it can damage pipes. But food waste in landfills causes methane emissions, so you shouldn't put it in the trash. Instead, look for ways to recycle and reuse it.

- Grease and fat can go into your compost with other food waste.
- Use it when you cook instead of buying processed cooking oils.
- Did you know you can make soap with grease? Look it up.
- A small amount of meat fat can be added to your pet's food to improve their coat.
- Grease and meat fat are a treat for some birds. **See Tip 64** on how to make a bird feeder.
- Some communities have drop-off locations for household cooking oil. Check with your recycling service or search online to locate where to take it.
- If you run a restaurant, make sure you use a recycling pick-up service for grease.
  - Grease has many large-scale reuses, like making biodiesel.

Tip 80 - Recycle single-use coffee pods. Send them back to the manufacturer.

When single-serve coffee makers became popular, it didn't take long for people to notice the amount of waste they generated. People complained, and companies started taking back the used pods to recycle them.

- If buying a single-serve coffee machine, find a brand that has a recycling program.
- Some manufacturers provide pre-paid envelopes to ship the used pods back to them and include a few in the box when you buy your new appliance.

Tip 81 - How to recycle electronics and light bulbs that can't go in your bin or trash.

Electronics and light bulbs are mixed materials that don't belong in your recycle bin. And they can't go in the trash because they contain chemicals that harm the earth.

Around 50 million tons of electronics are discarded every year, but only 20% of those are recycled. The good news is these materials are in high demand, so there are ways to recycle them, just not in your residential bin.

- Most communities hold regular e-waste events. They take items like computers, monitors, cables, and light bulbs.
  - Check with your MRF or local government for a schedule and specifics on what they take.
- Search for electronics recycling in your area. There are retailers that take used equipment, and many communities have collection sites. It's usually free, depending on the item.
- Retailers and e-waste events may not accept certain items like flat-screen TVs and incandescent light bulbs, the old non-CFL kind. Search online to find ways to recycle these.
  - There are mail-in programs available for fluorescent, compact fluorescent, halogen, and incandescent light bulbs.

Compact fluorescent lamps (CFLs) are great. We should all be using them since they last longer and save energy. But they're not so great for the planet. They contain mercury, which is highly toxic.

- Many retailers and hardware stores have recycling programs for CFL bulbs.

Planned obsolescence is when companies make devices that won't work after a few years. I discovered it with my first smartphone. The manufacturer didn't support software updates after it was a few years old. When I didn't upgrade and hung onto my paid-for phone, it soon malfunctioned. I was forced to get a new one.

Recycle electronics so we can reuse the materials to make more.

The need to buy a new device every few years adds toxic waste to our planet at an alarming rate. This is in addition to the staggering amount of greenhouse gases caused by making these electronics.

- Check with the manufacturer or the store where you bought the device to see if they will recycle it.

We need to be more responsible about reusing the materials in our electronics. We need to get better at recycling them. Mining new materials to keep up with our demand has proven damaging.

Electronics contain precious metals like copper, gold, and silver. Though in small amounts, not capturing and reusing these rare materials is often what causes product shortages.

Tip 82 - Disposable batteries aren't easy to recycle, but rechargeable ones are.

It's important we dispose of batteries responsibly. They can contain damaging chemicals that need to be recycled or discarded properly.

- They shouldn't go in recycling or trash. Batteries can cause a spark, start a fire, and injure workers.

We use various types of disposable and rechargeable batteries for things like remotes, toys, hearing aids, laptops, and cell phones. Each type requires a different recycling process.

- Rechargeable batteries can usually be recycled with other electronics. Disposable batteries can't.
    - Buy rechargeable batteries. They're eco-friendly, reduce waste, and are easier to recycle. They may cost more, but in the long run, they save money.
    - Single-use alkaline batteries are a lot harder to recycle and get expensive when you have to keep buying them.
    - Most rechargeable batteries, cell phones, and laptops can be recycled at stores and businesses that have recycling programs or at e-waste events.
- Very few places recycle disposable batteries.
    - Check out call2recycle.org. They accept all types of batteries through their mail-in program, including disposables. They also help you find where to recycle batteries and cell phones in your area.
    - Some waste management services, or MRFs, offer mail-in recycling of batteries.
- Button cell batteries, like from hearing aids, can be taken to electronics stores with recycling or an e-waste drop-off. There are also mail-in programs for these.
    - Save yourself time, money, and hassle; buy rechargeable button cell batteries. The manufacturer can tell you where to get them.

Tip 83 - How to recycle items like furniture, appliances, and household goods.

If you're getting new furniture, you're probably only thinking about how it's going to look in your home. But you also need to think about how to get rid of the old responsibly.

- If it's still usable, donate it. Many thrift stores offer free furniture pick-up.
    - Purple Heart and The Goodwill are two non-profits you may have in your area that pick up furnishings.

- o Give them away to someone who can use them.
  - You can sell or give away items within your community on Freecycle.org, Craigslist, Facebook Marketplace, or NextDoor.
- The materials in old mattresses can be used to make new ones.
  - o The Mattress Recycling Council offers mattress recycling. Go to byebyemattress.com.
- If items that contain metal no longer work, like appliances, metal framed furniture, bicycles, etc., there may be a metal recycler in your area that will pick them up.

Tip 84 - Motor oil and paint are dangerous chemicals that can't go in trash or recycling.

When chemicals like paint and motor oil end up in recycling, they prevent other items from being recycled and put the lives of workers at risk. But they can't go in the garbage since they're toxic.

- Your waste management service or local government should have information on how to dispose of these.
- There may be a special drop-off in your community for motor oil. This is the best option since these places will recycle it.
  - o Some auto parts stores and oil change places take used oil, some for free. They may also take the oil filter, which can be drained, and the metal recycled.
- You can't put leftover paint in the trash. Look for a paint store that might take it.

If you can't find a place to recycle motor oil and paint in your area, you'll have to take it to a hazardous waste drop-off.

What can't be recycled and why

Some items can't be recycled, and many can't go in the trash because they can be toxic.

Tip 85 - Dispose of chemicals correctly. They aren't recyclable, and most are toxic.

There are many ways chemicals find their way into our homes. Because they're dangerous, disposal is regulated and usually illegal to do any other way.

- Your cleaning supplies usually contain chemicals.
  - They are required to say on the label how to dispose of them properly.
- Always follow the guidelines for hazardous waste disposal provided by your government or waste management service.
- Use sites like Earth911.com or recyclenation.com to find where to take them.
- The US EPA's website provides information on how to properly handle and dispose of most chemicals.

Here are some chemicals people commonly need to dispose of and what to do with them.

- Automotive or engine fluids like antifreeze and transmission fluid are toxic and must be disposed of at a hazardous waste site.
  - Check with an auto parts store; they may know where you can take these.
- Paint thinners and other decorating chemicals, like wallpaper remover, are toxic and need to be taken to a hazardous waste drop-off.
  - Look for natural alternatives to these.
  - Rent a steamer to remove wallpaper rather than using chemicals.
- Cleaning products that aren't all natural can't go in your trash. These can be dangerous to people and the planet.
  - Check with your waste management service for what to do with them.
- Flammable liquids and chemicals, like lighter fluid, kerosene, nail polish and remover, pesticides, or items containing freon, need to be disposed of at a hazardous waste drop-off.

- ○ Check the Earth911.com database to find where to take them.
- Refrigerant Finders, refrigerantfinders.com, buys used and surplus coolants.

Tip 86 - Some paper products like napkins, paper towels, and tissue can't be recycled.

There are paper products that can't be recycled. Instead, they're made of shorter fibers or coated with plastic.

- Paper plates are made with short fibers that won't recycle into more paper. Look for alternatives and remove these from your shopping list.
- You can't recycle paper towels and napkins for the same reason, they're shorter fibers. Check **Tip 25** to learn how to reduce using these.

When in doubt, throw it out!

- Bathroom tissue, facial and toilet, are made of shorter fibers that can't be recycled.
  - ○ Use a hanky and roll back the toilet paper to reduce use.
- Paper cups - any beverage cup made of paper is not recyclable. They have a plastic coating that keeps them from dissolving when they get wet.
  - ○ No part of a carry-out drink cup can be recycled. Not the cup, the straw, or the lid. Even if it has a recycling triangle, it's still trash.
- Paper take-out food containers are coated with plastic and can't be recycled.
- Prescription bags are coated paper that can't be recycled.
- Freezer packaging has a plastic coating to keep it from melting. It won't get recycled in your bin.

If it's shiny or feels slick, it's coated paper. If put in your bin, it prevents other paper from being recycled.

Tip 87 - Greasy food containers, like pizza boxes, are bad for recycling.

A very small amount of grease, like a few splatters on the pizza box, will not contaminate recycling. However, greasy and oily residue can be a big problem.

- If the bottom of your pizza box is messy, but the top has no cheese or grease on it, cut the top off and put it in your recycling.
    - You may be able to toss the greasy bottom into your compost. Some communities allow greasy food boxes. Check with your curbside composting service, or throw it away.

Tip 88 - A container made of more than one material can't be recycled as is.

Any container made of multiple materials, like a plastic or metal rim on a paperboard container, can't be recycled as is.

- If you can safely take it apart, you may be able to recycle some of it, like the metal or paperboard but never the plastic.
    - There are instructions online for how to take apart a mixed material container safely.
    - If you can't get it apart, throw it out and look to buy the product in an eco-friendly container.

Tip 89 - Items made to be compostable can't go in recycling.

When things like compostable utensils came out, people wanted to buy them because they sounded ecological. Unfortunately, this is often false advertising. The company can say whatever they want on the product since compostable claims aren't regulated in most countries.

- Do your research and verify something is compostable before you buy it.
- Compostable items only go in the compost, not recycling.

Compostable doesn't mean recyclable.

The best thing is to stop buying single-use utensils and dishes. Chapter 2 has tips on how to reduce using these.

Tip 90 - Biological waste can't be recycled and needs to be disposed of properly.

Human and animal excrements – poop, snotty tissues, dead animals, and medical waste can't be recycled.

- Animal manure makes great fertilizer for your garden.
- Items like syringes cause contamination and risk to workers if put in recycling.
  - If you generate medical waste in your home, follow the instructions on the package for proper disposal.

Tip 91 - Small items we may think don't make a difference but do.

I used to not think about the little things I tossed in recycling. Instead, I believed some small items would get recycled, like store receipts, plastic caps, and the foil or cork from a wine bottle.

- Receipts belong in the trash. They're treated paper that prevents other paper from being recycled.
- Size really does matter... in your recycling bin. Even if foil can be recycled, the wrapper from a wine bottle is too small to go in the bin. Only bigger pieces of foil belong.
  - Sorting sensors can't detect small items. So they will end up in the trash.
  - If you think it can be recycled, look up where to take it or check with your MRF. Collect small items in a can to recycle when you have enough.
    - Only put foil with foil, metal caps with metal caps, etc.
    - Use a magnet to make sure something is metal.

Chapter 3 has ideas on how to reuse a number of small items like bottle caps and wine corks or look online to find other reuses.

If it fits in the palm of your hand, it's too small for recycling.

Here are a few other things you may think can be recycled that belong in your trash.

- Prescription bottles are too small to make it through sorting. Look for alternative recycling options for these.
- Cork can't be recycled.
    - If you have a lot of wine corks, find crafts and handy things to make with them. See Chapter 2 or look online for ideas.
- No loose lids or caps.
    - Plastic caps belong back on the container, so they make it through sorting.
    - Number 5 plastics are often tubs we buy food in that come with a lid. If your service says, they take 5s, put the lid back on the clean and dry tub before putting it in your bin. That way, the lid might get recycled.
- Gum wrappers are a mix of foil and paper that can't be recycled.

Your local recycling service or MRF is the first place to turn if you're not sure what to do with a disposable. If they don't take it, they may be able to tell you what to do with it.

Tip 92 - Rule of thumb: if it wasn't made to be disposable, it can't go in your recycling.

1-in-4 items put in residential recycling doesn't belong. These are usually materials that weren't made to be disposable. This is something we can fix.

- Laptops, cell phones, flashlights, furniture, dishes, cookware, etc., these types of non-disposable items frequently show up in recycling bins. They need to be taken someplace where they will get recycled or reused.
    - Residential recycling is meant for single-use materials. That's it!

o Things like electronics can be recycled, but not in a curbside bin.

There are companies that take back their products, like electronics. Most of these companies are being environmentally responsible, but not all. Some companies take back products only to make us think they're being responsible. This is called greenwashing. **See Tip 123**.

How We Got Here and Why We Can't Wait to Fix It

I heard a comment once that it's China's fault recycling no longer works. This is a great example of how limited knowledge leads to incorrect conclusions.

- With too many types of single-use plastics and how confusing it is, recycling was always going to fail.

In the 1980s, people in several countries, like the US, started buying cheap merchandise made in China and other developing countries. These manufacturing countries didn't have enough raw materials to meet the demand, so they started buying recycled materials.

More types of plastics were made, and contamination in recycling bins increased. As a result, the countries buying recycled materials discovered less of them could be used.

- Too many soft plastics and other contaminants were landing in our bins as single-stream recycling became the norm.
- How and what to recycle became confusing, causing most people to do it wrong.

Global waste grew as less of what China bought could be used to make products. This caused serious environmental problems for them. By this time, they had developed ways to mine their own materials, so they stopped buying our recycling.

- In 2018, China enacted Operation National Sword. Now the country only allows recycling shipments with less than 0.5% contamination.
- This caused a shock wave through waste management in the US, where recycling contamination is high.
  - China would no longer buy it, and there weren't many other buyers, especially since it couldn't be used for anything.

Many countries faced eye-opening realities when it became clear they could no longer ship their disposables off to another country and be done with them.

Out of sight, out of mind stopped working for our disposables.

Trash is a global issue we all need to face. A lot of waste is generated in our homes and at work. The only way to control it is to reduce how much trash we create and learn how to recycle correctly.

Imagine life without drive-thru and everyone carrying their lunches. Yep, that's where we need to be!

The wish-cycling problem and how to correct it

Wish-cycling is putting something in recycling with the *hope* it will get recycled. It has become a big problem, mostly due to plastics.

Wish-cycling is a curse that sabotages recycling.

In the 1990s, those of us who recycled in my community were conscientious about it. We paid attention so we would do it right. We had to; everything had to be clean and sorted.

Fast forward a few decades, and recycling is just *handled* by our waste management company, like our trash.

- Most people want to believe they're doing their part. They try and recycle but are frustrated now that our trash has become a staggering global problem.

Weren't our disposables supposed to be taken care of when we put them in recycling?

Tip 93 - We need to break the wish-cycling curse.

In one of my favorite futuristic detective novel series, everything gets tossed in a recycler. There are no trash bins. When I first read this, I thought, *please make this come true.*

Now that I know the truth about plastics, I'm afraid we are a long way from everything going into one bin. But this has already become a practice, and it is wreaking havoc on recycling and our planet.

It isn't our fault. We didn't create the system. But it's broken, and we must work together to fix it.

We must make recycling work. We must get smart about our trash.

- Take the time to learn how to recycle correctly everywhere you go; home, work, school, community centers, places of worship, etc., and help others learn.
- Eliminate the use of disposable materials that can't be recycled.
  - This is most plastics and mixed materials.
- Push your local government to control single-use packaging in retail and restaurants, especially plastics and foam.

Every community's recycling is different, and the items that are accepted vary. We need to learn about our local services if we're going to do this right, not go by what a friend says.

Investigate your recycling service, what they do with the materials you give them, and what they're doing to improve.

- Sometimes, they make changes due to new laws and regulations.
    o This happens when your vote makes recycling better.
- Tell your recycler or MRF you want them to do better. You're the customer. If you know they're not fixing a problem, tell them. They have contact info on their website.
    o Maybe they need to improve communications about what goes in your bin.
        ▪ Do they need better signs on bins?
        ▪ How good are their education programs to help you and your neighbors do it right?
- Check into Recyclebank.com. It's a rewards program. They help you learn how to recycle correctly and reward you for it.
    o Contact your MRF and encourage them to participate. There's a link on Recyclebank.com that says how.

The best cure for wish-cycling is knowledge. Learning how and why we recycle and what happens to those materials is how we're going to fix it.

It's on us to recycle right, even the tough stuff

Some of the things mentioned in this chapter are hard to recycle. But if there is a way, it's important we try.

Tip 94 - If your recycling service doesn't take it, don't give up.

If your service doesn't take an item, search for an alternative way to recycle it before throwing it out. Many hard-to-recycle items are accepted through mail-in programs or can be taken to a drop-off location.

Here are some free sites that can help you figure out how to do that.

- Earth911.com has an extensive database for recycling in North America. Search by item and your postal code to find where to take it. The site is full of articles and helpful resources.

- o I subscribe to their newsletter and listen to their podcast. Both are full of great information.
- Recycling Center Near Me, recyclingcenternear.me, where you can also search by item and location.
  - o They publish a blog with tips and insights on recycling and sustainability.
- Recyclenation.com has a large database of recycling resources in the US.
  - o Search on what to do with hard-to-recycle items.
- Recyclebank.com has a rewards program for correctly recycling.
  - o Publishes articles and tips to help you be a better recycler.

I won't lie. This is going to take some work. But first, we need to fix our recycling issues. That's going to take everyone doing it right and buying post-consumer recycled (PCR) materials.

We Need to Get This Right - Why Change is So Important

By now, you probably figured out that putting things in a recycling bin doesn't mean they get recycled. But why is that?

The main reasons are:

- Lack of education on how to correctly recycle,
- Not all recycling can be sold and turned into more packaging, and
- It's more expensive to make plastic packaging from recycled content than to use virgin materials derived from government-subsidized fossil fuels.

While I've talked about how Materials Recovery Facilities (MRFs) need to do better, some things are out of their control.

- Contamination in what they receive.
  - What we give them isn't recyclable, cleaned, or easy to sort.
    - Contaminated items either ends up as trash or are downcycled into disposables that can't be recycled again.
- What they can sell on the materials recovery market fluctuates.
  - Because of this, MRFs have a hard time staying in business.
  - Recycled plastics cost more than virgin, so the demand is low.
    - We need to increase demand for recycled plastics so prices will come down and more companies will use them.
    - Governments need to subsidize recycling, not fossil fuels.

To fix this, we need to find reusable alternatives and stick to what will get recycled. Chapters 1 - 3 can help.

Humans were unprepared for our modern lifestyle of waste

Because of messages from the plastics industry that told us recycling was the solution to our disposables, we weren't prepared for plastic waste.

Our disposable lifestyle has caused a lot of problems that can no longer be ignored.

- Over eight million tons of plastics go into oceans every year.
- More than a billion tons of trash goes into landfills annually.
  - Projected to be over two billion by 2025.
- Landfills emit methane, one of our most potent greenhouse gases.
  - Methane is 25 times worse than carbon dioxide (CO_2).

We were kept in the dark and didn't know most of our disposables were going into the earth and seas. Instead, we were told they were being recycled.

We aren't to blame - we were more duped into it than guilty of it.

Our waste habits crept up on us. Everything slowly became convenient and disposable. A lifestyle that was fed by a desire to believe in recycling.

Changing our trash problem may be a big task, but it's not hard. If everyone did these two things, we could change the course of our mounting waste.

- Significantly reduce the number of disposables we use and,
- Start buying recycled packaging to build a continuous loop of materials.

Buying materials that can be recycled, like paper, metal, glass, and number 1 and 2 plastics, is critical right now. This can prevent contamination and give us a better chance of reaching circularity with our packaging.

A closed-loop recycling system - what it is and why we need it

Closed-loop means disposable materials are continuously used to make more. This is what recycling means. It can happen with glass and paper, to some extent, and it happens with metals because they don't weaken and can be repeatedly used to make more. However, this can't happen with plastics because they degrade when broken down.

Plastics aren't easy to recycle, and most of them were never intended to be used more than once.

- There are numerous types of plastics, with more being made.

- o Most of them can't be recycled.
- o If they can be recycled, they can't be with other types of plastics.
  - ▪ A number 1 PET can *only* be recycled with other 1s. The same goes for number 2 HDPE.

The bottom line - plastics are what's wrong with recycling.

Recycling and reusing the materials we've already mined from the earth means we turn aluminum into more aluminum, paper into more paper, and metal into more metal. That is a closed-loop system. But it can't happen with plastics.

Though scientists are working on this, we're a long way from closed-loop recycling of plastics.

Tip 95 - Buy metal and glass containers and discover the ultimate in closed-loop.

Since plastics have less chance of being recycled, finding ways to stop using them is important. One way is to buy food and beverages in glass and metal containers.

- • Glass and metal have a better chance of getting recycled or reused.
  - o Effective bottle deposit laws can make this happen. If you don't have these, send an email to your local government representatives, and ask why not.

Metal is a hot commodity in materials recovery, so aluminum cans, or metal of any kind, can help your MRF be profitable. This can save you money on recycling rates.

- • Metal can be recycled over and over.
  - o 75% of all aluminum that's been mined is still in use.
- • Recycling aluminum uses 95% less energy than making new.

Glass isn't always easy to recycle, and it's heavy, so transport is costly.

- Emissions from recycling glass aren't much better than when making new.
- Different colors of glass can't always be recycled together, so sorting is a challenge.

Glass bottles work well in a reuse system.

Closed-loop doesn't always mean that materials are broken down, heated, or smashed to be recycled. It can mean reusing them *before* that happens. It means doing something with containers we should've been doing all along, refilling them.

- In Germany, they've had deposit laws, or schemes, for nearly twenty years. Once people got used to it, they reached an over 98% recycling rate. The scheme covers two types of bottles - reusable and single-use.
  - Reusables are returned to the bottlers, washed, and refilled.
    - Glass bottles are reused up to 50 times.
    - It's done at a local level to reduce transportation emissions.
    - Some Plastic PET bottles are included in the reuse program.
  - Most single-use plastic bottles are recycled into pellets used to make more plastics.
    - Over a third of Germany's plastics are recycled into more plastics. A higher rate than most other countries.
    - Packagers say they now use up to 70% less virgin plastics.
- Similar programs are showing up in the US.
  - Oregon introduced a refillable bottle deposit program in 2021. In certain areas, around 100 brands of beer, wine, and hard cider, are being sold in refillable bottles.
- Some retailers are participating in Loop, exploreloop.com, a global program that makes reusable packaging available in stores.
  - You buy products from a local store and return the containers to them when you're done.

We desperately need the reusable economy these programs promise, but they won't survive unless people start using them.

Tip 96 - Support bottle deposit laws; they improve recycling and create jobs.

Bottle deposit laws are the most effective way to recycle beverage containers.

The reason for this is simple. People return cans and bottles for the deposits they paid. What they return is usually washed, and food waste or non-recyclables don't get mixed in. Contamination is eliminated, so the only obstacle is getting consumers to return them.

- Ten states in the US have bottle deposit laws.
- Eight of those states have a bottle return rate of over 60%.
    - Of those, Maine, Michigan, and Oregon see over 80% of bottles returned for the deposit.
    - Deposit laws in some states aren't very effective because they make it difficult for people to return cans and bottles, like in California.
        - Bottle deposit systems need to be designed to encourage, not discourage, people to use them.

I lived in Michigan most of my life, where they've had a ten-cent deposit on beverage containers for over 30 years. Residents have gotten used to it, and the state now has an 89% return rate on cans and bottles.

- The law requires retailers to take back the beverage containers they sell.
    - Most grocery stores have machines where you insert bottles to get your money back. This makes it easy and convenient.

Bottle deposit laws and schemes create jobs and prevent the trash from building up. If you don't have these programs in your community, tell government officials that you want them.

Big Rewards - Recycling Done Right Benefits Everyone

People have said to me, "What's the point of recycling? I see my garbage people toss it in the same truck with the trash." That doesn't mean it isn't sorted at the facility. Some waste systems work this way. And some trucks are made with multiple compartments, so you don't see that trash and recycling go in different bins when picked up.

- If your MRF provides separate containers for your recycling, use them.
    - My MRF provides three bins, one for mixed paper, one for compost/organics, and one for all other recycling.
- Don't assume your recycling ends up with the garbage. Investigate your MRF. If their site doesn't say how they handle sorting, send them a message to ask.

I'm always afraid that people who see everything go into one truck will stop recycling. But we need to do it, no matter what, and we need to do it right. No matter how discouraging it may be, we must recycle as long as we have disposables in our lives.

Tip 97 - The more recycled packaging we buy, the less garbage we create.

We need to buy recycled packaging if we're going to develop a closed-loop system for our disposables. We need to develop a reuse mindset.

The whole purpose of putting materials in recycling is to use them again. To do that, we need to have post-consumer recycled (PCR) packaging on store shelves. Otherwise, putting stuff in our bin is pointless.

- Manufacturers need to use the materials we put in our bins to make packaging.

- Consumers need to be able to buy PCR packaging.
    - Research your favorite brands to see if they use recycled packaging.
        - If they don't, send them a message that you want it.
        Hint: write a standard message you can copy and paste into a company's contact form to make it easy.

Our buying habits drive demand for disposables and plastics, so it's our buying habits that will create a demand for PCR packaging.

- The more we buy, the more we will have available.
- Demand for recycled plastics will increase, and MRFs will be able to sell them.

Only consumers can shut down the one-way road our garbage is on.

Taxpayer dollars have been paying to manage the waste generated by corporations for too long. It's time we hold them accountable for their actions. In addition to reducing single-use packaging, we need to force companies to use recycled materials.

- Manufacturers need to take responsibility for the packaging they create, from how they make it to how it's disposed of.
- We need regulations like Extended Producer Responsibility (EPR) laws.

Key to having companies use recycled materials are laws and regulations that require it.

- EPR laws mean building infrastructure to recycle packaging. This means jobs.

Recycling Creates Jobs and Boosts the Economy

We need facilities to recycle the materials we've already mined from the earth. This is going to take a lot of work and a lot of people. It turns out that saving our planet can also save our economy.

A circular economy is a sustainable one.

Building recycling infrastructure is not only critical to saving the planet but can also be an economic windfall for communities that embrace it.

Tip 98 - There are plenty of jobs and opportunities available in the recycling industry.

The recycling industry currently provides over half-a-million jobs in the US alone.

- The Institute of Scrap Recycling Industries, ISRI.org, reports that recycling contributes almost $117 billion to the US economy each year.
- Recycling generates $5 billion a year in revenue for local economies.

Learn about opportunities in recycling and how they benefit the economy by visiting ISRI.org. Their site includes training and information on careers in the industry.

Building family and community - let's save the world together

There are always ways to learn how to recycle better, and it's one of the healthiest family activities you can do. Being involved in your community recycling programs is a great way to make sure disposable materials don't end up as trash.

- Go to your local government website and search on recycling or sustainability.

- - If they don't have resources and education tools, ask them why.
- Educate yourself and teach others how to get better at recycling.
  - Many websites for MRFs and governments have a children's section to teach kids about recycling and why it's important. See Chapter 7 for more ideas.

Tip 99 - Recycling must be a group effort if we're going to make it work.

Putting disposables in recycling bins happens in our homes and our communities, so that's one place it needs to be fixed. We can get it right if we work with each other and our recycling services or MRFs.

Here are some ideas to help you, your family, friends, and neighbors get involved.

- Ask people what they know about recycling or why they think it doesn't work.
  - Listen to them, then look up what they said to find supporting data and maybe a solution before you repeat it.
  - Know the facts. Misinformation can hurt recycling and make it confusing.
    - Then people do it wrong or care less about doing it right.
- Challenge your friends, family, or neighbors to a *trash down*. Create challenges around who generates less trash and recycling.
  - Keep track by marking your calendar or having a weigh-in with neighbors.
    - The household that takes out less trash in a week or month wins.
  - Get creative and see what challenges you can come up with.
- Don't stop your green living habits just because you're entertaining. Practice recycling and zero waste when you have people over. **See Tip 36.**

- o Friends and family may pick up planet-friendly habits by watching you.
- Help kids to understand the importance of recycling the right way.
  - o Ask them if they have ideas on how to reduce trash.
  - o Maybe they can make it a school project and get credit for it.

There are plenty of resources to help us get recycling right. Beyond what we can learn in our communities, there are national and global programs to reduce waste and recycle correctly.

- Global Recycling Day, globalrecyclingday.com, usually happens yearly in March.
  - o They promote global efforts to educate and build good recycling habits and infrastructure in our communities.
- The Earth Day organization, earthday.org, is active more than one day a year.
  - o They have loads of resources to help us learn how to care for the planet every day.
- Subscribe to newsletters that keep you updated on recycling and what's working and what's not. Your local government or MRF may have one.
- See the resources listed in **Tip 94**.

Everyone wins when we do our part to recycle right

There are many perspectives on recycling. Some people see it as an added expense and not worth the hassle. Since we know that isn't true, and it's the best solution to our disposable lifestyles, we need to build the habit of doing it right.

Spend a little time each week to make sure you know how to recycle correctly. This is a great way to fight climate anxiety and feel positive about the future.

Tip 100 - If you run a business, you have control over your recycling practices.

Most people spend their waking hours at work, where they constantly use disposables. If you own a company, you likely have policies and standards for your employees and customers. Make sure these include recycling practices.

- Have paper recycling bins near printers.
- Install recycling in customer waiting areas, kitchens, and employee break rooms.
  - Put up signs to tell people what goes in them.
- Make sure your janitorial service properly recycles what is put in bins and doesn't throw it out with the trash.
- Include recycling and sustainability policies in employee manuals.
- Have a recycling or sustainability message in employee newsletters.
- Post Reduce, Reuse, and Recycling tips in common areas.
  - Put up a new one each week so it's fresh and people look at it.
- Sponsor and organize volunteer activities like local clean-up events.
  - Put together a green team to organize these.
- Encourage your staff to have an eco-friendly message in their email signature, such as, "Think of the planet before printing this email."
- If you are in manufacturing, make sure excess materials are recycled.
  - Reducing waste will save you money.
  - **See Tip 147** for organizations that can help.

There are many resources online to help businesses be more sustainable. Don't make it a one-time thing. Make it part of your mission statement and use the above steps, so it becomes part of everyday operations. It will help your bottom line.

- It improves public relations when you have recycling and sustainability practices.

- People want to buy from and do business with sustainable companies.

You don't have to own a business to encourage recycling practices at work. Talk to your employer if they aren't doing the things listed above.

One of the biggest impacts we can make is to change how we treat our trash. We won't get it under control until we figure out how to recycle the right way and reduce what can't be recycled.

- It's a waste of time, energy, and resources to put materials in recycling bins only to have them end up in landfills or oceans.

Here are the most important takeaways I hope you gained from this chapter.

1. We've been recycling for decades, but it hasn't worked because it wasn't set up correctly or properly funded.
   a. It's not our fault. We were led to believe it worked, so we would keep buying disposables.
   b. Corporations, the fossil fuel industry, and the makers of packaging need to be held accountable. We need laws to make them correct the broken system they've created.
2. The recycling triangle doesn't mean much.
   a. It's not regulated and shows up on items that can't be recycled.
3. Services that collect recycling are Materials Recovery Facilities (MRFs) that only sort and sell the materials. They don't recycle them.
   a. They can't sell them if manufacturers aren't buying them.
   b. There isn't a market for recycled plastics because virgin plastics that come from government-subsidized fossil fuels are a lot cheaper.
4. What your recycling service, or MRF, can accept is constantly changing.
   a. Regularly check their website to avoid contaminating good materials that are recyclable.

5. Wish cycling is the practice of putting things in recycling with the *hope* they will get recycled.
    a. It is very damaging because it prevents good items from being recycled.
    b. To prevent it, only put in recycling the materials you know are accepted. *When in doubt, throw it out.*
6. The goal of recycling is to create a closed loop of disposables so we no longer mine the earth for virgin materials. This only happens when what we already have is used over and over.
    a. This is the purpose of recycling, but it isn't working yet.
    b. We need to buy recycled packaging to create this circularity.
    c. Government regulations and incentives are needed to make this packaging available and reach the goal of a closed loop.
7. Building better recycling programs to repurpose materials we already have creates jobs and boosts local economies.

Chapter 5: Undoing the Damage - Reversing Our Impact on the Planet

Greenhouse gas (GHG) emissions are now at the highest levels since the beginning of human existence. The results are severe environmental changes that we have little choice but to prepare for and hopefully reverse, or many of Earth's species, including humans, will continue to be threatened and become extinct.

Our collective activities cause GHGs; Our collective efforts are needed to reduce them.

Reducing emissions and restoring the earth's natural systems are critical if we're to stop the warming of the planet. To accomplish this, we need to look at the source of GHGs.

Our two most damaging GHGs are carbon dioxide (CO_2) and methane. The most damaging is methane, which scientists say must be reduced to stop rising temperatures.

- Methane traps a hundred times more heat than CO_2 over a five-year period.
- There is two and a half times more methane in our atmosphere than there was before industrialization.
- More than 60% of methane emissions are caused by human activity. The worst being...
  - Agriculture and livestock farming - cows are the biggest source.
  - Landfills - food waste causes most of the methane from this source.
  - Fossil fuels - unless we're going to stop driving cars and heating our homes, a transition to renewable energy is crucial.

While methane may be the most damaging, there is a lot more CO_2 in our atmosphere, so it's just as harmful.

All life is carbon-based. To support it, we need a proper balance of CO_2 between the atmosphere and the surface. As we lose this balance, our planet becomes less habitable. This starts to happen when CO_2 levels go over 350 parts per million. In April of 2021, NASA recorded CO_2 at 416 ppm.

- The biggest source of human-created CO_2 comes from burning fossil fuels.
- The US has the highest CO_2 emissions per person than any other country.

If we recognize that what we do has an impact, we have a chance of fighting climate change.

We need to stop thinking that small amounts of waste are insignificant. Every bit makes a difference, from the trash we generate to the water and energy we waste.

Land, Air, Sea - How We Can Help Them Heal

Like many of you, I learned about air and ocean currents in grade school–how the flow of these currents maintains a livable planet. But unfortunately, we have disrupted those currents.

Though seasons continue and deniers will say global warming doesn't exist because it snowed on them today, the disruption in our currents is undeniable. It's why we see heat waves in the Antarctic and ice storms in Texas. And it's why rain patterns are no longer consistent, causing droughts and floods.

Human actions got us into this mess and are the only thing that can get us out.

It took decades of human activity to disrupt our currents, so it's going to take time to fix them. The weather disruptions will continue, but if we stop doing what caused them, we may be able to reverse the damage.

Like our bodies, if we start eating healthy, we won't feel better tomorrow. But if we continue taking better care of ourselves, we eventually get the reward of good health and a longer life. Earth is no different.

Healing our planet - there's more to do than plant trees

We need to help our forests and oceans to heal, but this can be a daunting task that feels out of our hands.

- Big corporations cut down carbon-capturing forests.
- Manufacturers pollute the earth with toxic chemicals and GHGs.
- Fossil fuel companies kill ocean life by drilling for oil and gas.

The ugly truth is they do these things because we pay them to.

Tip 101 - Eating these two foods is the biggest cause of deforestation.

Deforestation is one of the most damaging things we do to our planet. It causes poor air quality and soil erosion, destroys animal habitats, and puts more CO_2 into our atmosphere than all the cars on the planet.

- In 2021, it is estimated that we lost woodlands the size of a football field every second.

The best way to fight deforestation is to pay attention to where our consumer demands cause it.

I used to think trees were cut down to make paper, but that is not the biggest reason. While we really need to cut out using paper convenience items, **see Tip 25**, it's more important that we stop eating foods that require a lot of land to produce.

- **Our demand for beef** causes a double whammy effect on our planet since it increases CO_2 *and* methane emissions.
  - This deforestation is responsible for 15% of our carbon emissions.
    - Rainforests and tree groves are cut down or burned to create grazing land and feed crops for cattle.
    - Land is cleared to grow soybeans, of which humans consume only 5%. The rest is used as a protein supplement to feed livestock.
  - Cows are the biggest agricultural source of methane. Gas in a cow's stomach causes methane when they burp. It also comes from their manure.
    - With 1.3 billion cows on Earth, that adds up to a lot of methane.
- **Palm oil is in 50% of the products we buy.** From processed foods to makeup, it's heavily used because it's one of the cheapest vegetable oils to produce.
  - Palm oil plantations in Indonesia have killed nearly 25 million acres of forest in the last 20 years.
    - 80% of Indonesia's forest fires are set on purpose to clear land for palm plants.
      - Burning trees emit more CO_2 than cutting them down.
    - Indonesia is a small country but is among the top five in forest loss.
  - Our high demand for palm oil *must stop*. But this is challenging because it's in a lot of products and isn't always identified.
    - It can be an ingredient *within* a listed ingredient. Sometimes we don't know we're buying it.
    - We must do research to find out where it is and avoid it.

Our largest natural collectors of carbon are rainforests. Yet, over the past few decades, more than half of them in Brazil have been lost to livestock farming and logging. This was slowing down due to preventative laws in the country, but new leadership has defunded the enforcement of those laws, and deforestation is on the rise again.

- Amazon Watch is a non-profit fighting to protect the Amazon Rainforests. They run campaigns like #AmazonCeaseFire to fight this continued destruction.

Since we can't rely on governments and businesses to do what is right, we must take matters into our own hands.

- Stop eating beef. If it's something you do regularly, only do it once a week and work up to once a month. If you don't do it often, try cutting it out completely.
- Read product labels and find alternatives to buying anything containing palm oil.
  - Stay clear of processed foods that use palm oil as a preservative.
- Eat more plant-based foods.
  - Designate one day a week to eating only plant-based or consider becoming vegetarian or vegan.

The bottom line, it's a lot healthier for all of us to eat less beef and stop buying anything that contains palm oil.

Tip 102 - Support regenerative farming to heal the planet and protect our food supply.

Eating more plant-based foods is good for the planet, especially when those plants are grown sustainably and with regenerative farming practices.

Regenerative means non-invasive practices that work with the natural soil composition.

To produce more food, agriculture companies use chemicals and do things that damage the soil and our ecosystem.

- They use fertilizers, pesticides, and herbicides to get the most out of every crop.
  - These chemicals end up in our food supply, groundwater, and aquifers.
- With big machines, they till the dirt releasing carbon.

- Deep tilling or turning the soil is bad for plants and the environment.
  - Tilling releases carbon and organisms meant to help the soil fight off weeds and pests naturally.
  - Tilling reduces the vitamins and minerals we expect from our produce.
- A standard farming practice is to clear picked fields, often burning the plant remnants, which increases CO_2 in the air.
  - Letting the remaining growth compost back into the soil is how vitamins and nutrients develop.
    - This is regenerative farming and the way nature works.
- Healthy soil organically controls pests and weeds.
  - This requires regenerative practices, like rotating crops and fertilizing with organic matter like compost and manure.

Sustainable and regenerative farming means working with the soil and its natural composition so it's healthy and chemicals aren't needed. It means growing food the way nature intended.

- Planting the right crops and rotating them.
- Growing cover crops that don't produce food but support healthy soil that will one day produce better food.
- Not using chemicals and building up healthy soil that does what it's meant to do; protect and nourish plants.

Buying locally grown organic produce is healthier for you and supports regenerative farming. You can still support this when you buy produce at a grocer. Here's how to look for organically grown.

- Check the UPC label (the little sticker with a bar code). Produce grown by a certified organic farm will have a UPC code that starts with a 9.
  - The International Federation for Produce Standards (IFPS) monitors and regulates this labeling.

Tip 103 - Grow your own food and practice regenerative gardening for healthy soil.

Big commercial growers aren't the only ones who need to practice regenerative and sustainable growing.

Regenerative gardening does many things to help combat climate change.

- Heals topsoil, so it absorbs carbon and promotes healthy growth.
- Increases biodiversity that supports life and our ecosystem.
- Improves the water table so we can grow more food in the future.
- Provides biosequestration of carbon dioxide (CO_2).

You save money, eat healthier, and help reduce global warming by growing your own food. You don't need a yard. A few canisters will do.

- You don't need a lot of space to grow your own food.
  - o With a small patch of land, you can grow a few plants.
- Canister gardening is great for apartment living or if you don't have a big yard. All you need is sunlight.
  - o Having house plants is good for your health and improves air quality in your home.
- Homegrown produce is healthier.
  - o Vitamins and nutrients develop in fruits and vegetables during ripening.
    - Commercial growers pick produce before it ripens, so we lose most of the health benefits when it's shipped to us.
  - o Organic homegrown or locally grown tastes better because it ripens on the vine.

Regenerative gardening will make the soil more productive, promote healthier growth, and support a sustainable planet. Here are some basics you need to know.

- Don't use pesticides or weed killers. There are natural ways to keep out bugs and unwanted plants.
  - Check out diatomaceous earth. A natural compound that is safe for human consumption and keeps most bugs off plants.
    - The powder disables them by getting stuck to their legs and wings.
  - When soil is at its healthiest, it has the organisms and microbes to fight off pests and weeds.
    - Pulling weeds is better than using chemicals that kill them *and* the soil.
- Practice biodiversity with your crops. Research plants that complement each other.
  - Look for ones that will heal the soil and regenerate the nutrients needed for healthy growth.
- Cover your soil with mulch to keep in moisture which helps during droughts and prevents erosion in heavy rain. This helps the soil retain nutrients and prevents carbon from escaping.
  - Don't buy mulch. Instead, grow cover crops that provide mulch for your soil or use leaves you collect in the fall.
- Don't turn or till the dirt beyond a few inches of topsoil. Learn the best ways to prepare your soil for planting.
  - Study up on regenerative gardening. Check out Green America's Climate Victory Gardens for tips.

If we want a healthy planet, we need to stop mass deforestation and support organic farming. If we cut down fewer trees, replace what is cut down, and grow food the way nature intended, we have a better chance of reversing global warming.

We have the power to stop polluting the air we breathe

You likely already know that our planet is not healthy. In fact, the way I see it, we are currently on life support, and the prognosis is not good. But we can change this.

Humans are smart, and we have the power to undo the damage. That power starts with knowledge.

It wasn't until I was in my thirties and worked in the energy industry that I discovered that my electricity came from burning coal. In some regions, it comes from natural gas. Extracting and burning these fossil fuels is the biggest cause of climate change.

Around three-quarters of our energy is provided by fossil fuels.

The good news is we're ready for change. Scientists warned us a long time ago, so several smart people have been hard at work preparing for the transition.

- Global capacity for renewable energy increased by 45% in 2020.

It's time to let fossil fuel companies know that we want to buy renewables. Then, they can change how they provide us power and still be profitable. Investing in renewables is just a different energy business, especially as more subsidies become available.

- We will always be buying energy, no matter the source.
- Our energy providers need to be told that we want renewable energy.
  - o If energy providers make the transition to renewables, the prognosis for life on our planet will quickly improve.
    - Many providers have started to invest in wind and solar technologies, but this isn't happening fast enough.
  - o Governments need to stop giving them our money for incentives to keep burning fossil fuels.

Energy companies aren't the only ones who need to change. Our individual lifestyles need to support the transition.

Humans have developed many habits and luxuries that contribute to our climate crisis. However, a number of these can be easily changed if we pay attention and try.

Tip 104 - Give up your love for gas cooking and appliances.

More potent to climate change than carbon dioxide (CO2) is methane gas. Natural gas is one of the biggest contributors to methane. Production in the US alone contributed nearly 20% of methane emissions in 2020.

- Replace gas appliances with electric.
  - Use an electric grill or convert your existing gas grill. This can be done with certain grills. Look it up.
- Get an electric heat pump. They use thermal energy to heat and cool a home or building.
  - An economical and easy way to make your home energy efficient.
    - Look online for incentives and rebates in your area.
    - There may be tax credits available.

Tip 105 - Gas lawnmowers and small-engine vehicles are big polluters.

I grew up in a state with a lot of lakes and water sports. My dad owned a boat, and I knew many people with gas-powered toys on lakes. We didn't know any better. Now that we are feeling the effects of those emissions, we recognize the need to find ways to entertain ourselves that won't harm the planet.

- Gas-powered lawnmowers, all-terrain vehicles, and watercraft contribute to our carbon emissions at shocking levels.
  - In California, studies showed that off-road gas engines cause more emissions than all the cars in the state.
    - This prompted a law that will ban the sale of small gas-powered engines starting in January of 2024.

- If you are looking to buy a small engine off-road vehicle, lawn mower, generator, or chainsaw, look for zero-emission or electric alternatives to gas.

These can mean big lifestyle changes for some people but are invaluable to the future health of our planet and our children.

Some things are easier to do because they simply mean tweaking our current activities, like how we use energy in our homes.

Tip 106 - Unplug, power down, and send fewer texts and emails to reduce energy waste.

Energy consumption has grown exponentially since the dawn of the digital age. We're plugging in and connecting more than we ever have. The challenge is how much of it we waste.

Around one-fourth of generated energy is wasted.

Eliminating this waste will reduce coal and gas production rather than increase it, which is currently happening.

- Even after global leaders have agreed to take action to stop drilling and mining, these are still growing industries.

Things that use little energy seem insignificant when they aren't. Having unused power cords plugged in, leaving lights on, or even sending emails or texts are small actions that add up to a lot of energy use when we consider how many people are doing it.

- Send fewer texts and emails. Again, it seems like a small thing, but it's not. We don't think about it because the energy used is so minor, or we may not realize how it uses energy. I know I didn't.
  - Every text or email we send is stored in a data center somewhere. This increases our energy use and impact on the environment.
    - Some data centers use enough energy to power 80,000 homes. The more messages we send, the more this energy use will grow.

- Pay attention to *vampire devices* that suck energy when not in use. Things like phone chargers and small appliances account for a lot of wasted energy when we add up how many of them are left plugged in.
  - Unplug electronics and small appliances like computers, game consoles, blenders, and toasters— especially ones you don't use regularly.
    - Phone chargers are big culprits because so many are left plugged in when not in use.
- Always turn out lights that aren't needed or when you leave an empty room.
- Don't leave bathroom appliances plugged in, like blow dryers or razors.
- Don't leave your computer on; shut down, unplug it, or turn off the power strip every night.
- Use energy-saving power strips to make it easier to shut things down.
- Unplug everything you can when going away for more than a day, like the microwave, TV, or entertainment center.
  - Use power strips to shut these down, so it's easier.
- Turn on *energy-saving* mode for any appliances with this setting.
- Turn off *quick start* or *instant on* settings for your electronics and TVs.
  - These convenience settings drain energy even in power-save mode.
- Put a lid on a pan when boiling water - it boils quicker and saves energy.
  - Don't overfill the tea kettle or hot water pot. Only heat the water you need.
- Keep your freezer full. Empty space requires more energy to freeze food.
  - I don't usually have a freezer full of food, so I freeze water bottles. Then I'm prepared for power outages, and I don't have to buy ice for coolers when traveling.

Why not start a healthy routine of unplugging, taking a twenty-minute meditation break, or going for a walk? Your body will thank you, and so will the planet.

Tip 107 - Make your home energy efficient to help the planet and save money.

Renting or owning, no matter where I live, I always look to seal leaks around doors and windows. In addition, I do what I can to make sure my home is energy efficient, not just because it saves money but because it reduces my energy use.

There's more to do than covering cracks around windows. Here are some ways that may work for you.

- Service providers help make your home more energy efficient, and often for free. They need us to reduce our energy demand on the grid.
  - Your provider may have incentives for energy-saving devices.
  - They may know about rebates and deals on energy-efficient appliances, like electric heat pumps.
- Many energy providers offer lower rates when you agree to take actions that reduce your energy use.
  - Look into a smart thermostat program. You can likely get one installed for free or at a discount. These save energy and money.
- Energy companies offer online services for you to monitor your energy use. Your provider may even have an app for this.
  - You can monitor how the changes you make reduce your energy use and your bills.

Many governments offer incentives that make it easy and affordable to upgrade and make your home more energy efficient.

- In the US, the recently passed Inflation Reduction Act includes energy reduction incentives for homeowners.

Tip 108 - Get an electric or hybrid vehicle to save money and the planet.

Owning an electric vehicle (EV) has gotten easier and less expensive, and hybrid EVs can make it less difficult to embrace the transition. And faster-charging technology is reducing the hassle of owning an EV.

- Many communities are adding charging stations in public places.
- There are often incentives that subsidize the purchase of EVs.
  - Search online or ask your local dealer.
- Tax credits are usually available to those who purchase an electric or hybrid car.
  - Incentives could be available to help you install a charging station at home, which adds value to your property.

Don't discount the opportunity to go electric just because you checked once and it was too expensive or inconvenient. New technology, incentives, and increased demand are making this easier.

- Worried that charging will be too difficult and time-consuming? Consider buying a hybrid vehicle.
  - A plug-in hybrid will run on gas when you aren't able to charge.
  - A full hybrid will run on both gas and electric without having to plug in.

Tip 109 - Walk, bike, carpool, and use public transportation to reduce emissions.

The communities with the best air quality are the ones that are bike friendly. They encourage people to ride bikes to work and school with safe bike lanes, secure places to park, and monetary incentives.

Vehicles on roads account for one-fifth of greenhouse gas emissions; around 40% of this is from passenger vehicles.

This is something we have control over, and it goes beyond how much we use our cars. The emissions that don't come from driving our cars come from shipping goods to our doorstep or neighborhood. But first, let's look at how we can reduce emissions from our personal vehicles.

- Bikes are often used for exercise, but why not for short trips or to work?
  - Need a few items from the store? Why not take the bike and a backpack?
  - Bike together as a family to a local restaurant to eat out.
- Join a carpool. There are apps to help you connect with other commuters who make similar drives.
  - What a great way to make new friends and connect with other environmentally conscious people. Plus, you'll save money.
- Public transportation is well maintained and reliable in many urban areas, as well as some non-urban communities.
  - Combine riding your bike with taking the bus or subway to work or school.
  - Though some can be unreliable, it's worth investigating how well your transit services work.
    - The more we use it, the more we help improve public transit.
  - You may find you have time to read, and you'll be much more relaxed after your commute.

New climate initiatives include building up public transportation and making it easier to use. If it's available in your area, try it. You may like it.

Tip 110 - Rethink your personal travel and how you vacation when planning a trip.

You work hard and need a break. When it's time for relaxation, try and do it in ways that are less harmful to the planet.

We know that the biggest cause of CO_2 emissions is transportation, so how we handle our personal travel makes a difference.

Vacations should cause less damage to Mother Earth, not more.

Choose ground transportation whenever possible, but when you do travel by air, there are things you can do to reduce the impact. While emissions from air travel cause less than 10% of greenhouse gases, flights for tourism cause nearly 80% of travel industry emissions.

- Avoid short flights and book direct whenever possible. Most emissions happen during take-off when the plane is lifting the weight it's carrying.
- Pack less whenever possible to reduce weight on the airplane.
  - I have a bad habit of packing clothes I don't end up wearing or books I don't read. I never thought about how I was adding to the weight of the plane and, therefore, my carbon footprint.
- Carry a refillable water bottle. You can go through security with it empty. Most airports have water fountains made for filling these.
- BYOF - Bring your own food. It's cheaper and more ecological than buying food at the airport, which is usually packaged or served in disposable containers.
  - Carry your own snacks and pre-made meals in reusable containers or baggies.
- Consolidate visits to family and friends in the same region. Rent an electric or hybrid car and drive around to visit them.
- Road trips are a great bonding time for families. Reconsider flying to your next vacation spot and make it a road trip instead.
  - There is so much you can learn about the world and each other on the open road.

Tip 111 - Be an informed consumer to reduce emissions when you shop.

Invisible to most of us is how our increased consumerism has upped the amount of energy wasted. I'm not saying that wasted energy is entirely on us. Not by a long shot. Corporations and manufacturing waste more energy than individuals. But they make their money from us, so we have a say in how they do business.

The amount of energy wasted in our homes is nothing compared to that of businesses. The biggest energy waste comes from...

➤ Transportation

➤ Manufacturing

➤ Grocery stores.

Companies in these industries must become energy efficient. We encourage this when we buy from companies that act responsibly. From the goods we buy to the places we eat, we need to do business with responsible companies that work to reduce emissions and waste.

- Research your favorite brands; if one isn't sustainable, look for a new favorite.
    - Read their sustainability or corporate and social responsibility (CSR) reports. Do they give details, or do they use fancy words to sound sustainable without giving facts? This is called greenwashing. **See Tip 123**.
    - Check their energy, water, and chemical policies. What do they do to reduce waste or make these eco-friendly?
    - **See Tip 127** to learn how to shop for sustainable products.
- Buy local and used goods to eliminate manufacturing and transportation emissions.
- Look at your local grocery stores and what they do to reduce emissions and energy waste, like sealing refrigeration and moving away from HFC coolants. **See Tip 117**.

When we avoid buying new, we reduce energy use and the emissions caused by manufacturing.

Our lives depend on us saving our oceans

Oceans sustain life through an intricate ecosystem balance and the oxygen they provide.

- As much as 80% of our oxygen is generated by marine life.
- Healthy oceans are needed to absorb CO_2 and reduce it in our atmosphere.

Coral reefs are the backbone of oceans. They're like giant cogs in the whole machine of life and necessary for the survival of a biodiverse ecosystem. So we need coral to maintain marine life and life on the planet as we know it.

- At coral's current rate of demise, it is estimated that as much as 90% could be dead and gone by 2050.

It's clear we cannot survive without our oceans. From the smallest plankton and algae that live on the ocean floors to the large mammals living near the surface, we depend on them all.

Tip 112 - Do your part to save the oceans and sea life that support our lives.

There are several organizations that offer opportunities for you to pitch in and help save our oceans and sea life.

- Find ways to support ocean-saving efforts online or in your community.
- Donate time or money. There's always something you can do.

Here are just a few resources where you can find ways to do this.

- **Parley for the Oceans**, www.parley.tv, offers steps you can take to become part of the solution. They have events, educational materials, and ways to get involved.
- **Ocean Conservancy**, oceanconservancy.org, reports on threats to ocean life and how we can stop them. They'll help you stay informed about how to save oceans when you go to vote, and they sponsor coastal clean-up events.

- **US National Ocean Resource** programs are a great place to learn. Go to oceanservice.noaa.gov and visit the Education Resources link to get the latest Ocean Service Education materials sent to your inbox. Great for helping kids learn why oceans are so important.
- **Oceana.org** provides the latest news on how to protect our oceans. They have ways to send messages to your representatives about actions that need government support.
- **Ocean Blue Project**, oceanblueproject.org, gives resources for creating or joining beach clean-up events in your area.
- **Mission Blue**, mission-blue.org, supports campaigns and actions to save our oceans. They promote public awareness and support for marine protected areas.

SOS/Save Our Seas = Save Our Souls

Most people have heard of the Great Pacific Garbage Patch, aka Plastic Island. What many don't realize, at least I didn't, is that's only the biggest island of garbage. There are now five giant plastic islands, along with a lot of little ones. They are formed by ocean currents that bring floating plastics into their whirlpool.

At the rate we're going, plastics will outweigh fish in the oceans by 2050.

The scary part is the plastic in these islands is only 15% of what is in our oceans.

- The Ocean Clean-up, theoceancleanup.com, is a non-profit that recovers ocean plastics and makes new items from them.
    - They sold the first sunglasses made with plastic from the Great Pacific Garbage Patch. Visit their site to learn more about how you can help to reduce ocean plastics.

The islands of garbage in our oceans are growing exponentially. Studies have shown that the Great Pacific Garbage Patch, which is currently the size of Texas, will grow tenfold in the next few decades if we don't change what causes it, our use of disposable plastics.

- Plastic fishing nets are dumped in oceans regularly and hold together these islands of plastic.
- Many consumer-rich countries have shipped contaminated recyclables to underdeveloped countries, where they're found to be useless and often dumped in oceans.

Tip 113 - Don't buy things that cause chemicals to end up in oceans.

Another major cause of death in our oceans is toxins that leak into waterways or are dumped there. Chemicals are used a lot in manufacturing; two of the worst are agriculture and the apparel industry.

- Pesticide used for agriculture is one of the worst things for our oceans and for people. These chemicals spill into our waterways and seep into the groundwater we drink.
  - When we support organic farming, we prevent pesticides and other chemicals from entering our drinking water. **See Tip 102**.
- While colorful clothing may be fun, it's not so fun for our ocean friends or our bodies. Most apparel is colored with harmful chemicals that we need to avoid.
  - After chemicals are used to color clothes, what's left is dumped into the ground or waterways. **See Tip 128**.

We need to give our oceans a chance to heal. We need to stop poisoning them and life on the planet.

Once our planet is healthier, we will be healthier.

I read a report recently that said the climate crisis is a bigger threat to human existence than Covid-19. This isn't a surprise considering the World Health Organization reports that the health of millions of people is negatively affected by poor air quality every year.

There are many things that have contributed to our unhealthy environment. It's not just the emissions in our atmosphere or the plastics choking our oceans. And it's not only our disposable lifestyles that caused our current climate crisis.

It's taken combined human activities to get us where we are, so it will take combined efforts to fix the problem.

Visions of Tomorrow - Where Will We Be, What We Can Do

Human activities have caused over a one-degree Celsius increase in average global temperatures. When I first heard that an increase of three degrees would make the planet uninhabitable, I thought it strange that two more degrees could make that big of a difference. Then I learned that this is like the Richter scale for earthquakes; one-tenth of a degree increase causes exponential damage.

- This damage is showing up in the climate catastrophes we're now seeing globally.
  - Fires, floods, droughts, and climate refugees are all on the rise. We can't look at the news and not read about them.
- On our current path, scientists say a 1.5°C increase is inevitable. But it's preventable and even reversible.

The science is clear; there will be serious repercussions if we continue doing things the way we've been doing them.

- At a 1.5°C temperature increase, over 70% of coral reefs will be dead.
  - On top of supporting our lives, coral reefs prevent coastal erosion, which is important as sea levels rise.
- With melting polar ice, sea levels are expected to rise by 56 centimeters, that's almost two feet.
  - Many coastal areas will be underwater, and the millions of people living there will become climate refugees.
  - In the US alone, over four million people currently live in coastal areas that are at risk of being underwater in the next decade.
  - Tropical islands, like Hawaii or the Maldives, are expected to disappear once we exceed 1.5°C warming.

- Heatwaves will continue to increase, creating droughts. This will leave us fewer livable areas and less land for farming.
  - The 21st century has seen more repeats of record-high temperatures than ever before.
  - We're losing fertile land to grow food. Scientists say some areas that are drying up will never recover; crops will never be grown there again.
- New studies reveal that many species are already at maximum livable temperatures.
  - It won't take much to wipe them out, and it won't happen gradually. Many species will just suddenly disappear.
  - The circle of life ensures that this will happen across multiple species at the same time.
    - Every living thing on Earth depends on another species in some way. So when one dies, many others will be affected.

The biggest ecosystem collapse may be around the corner

All the things we need and care about are diminishing - bees, flowers, trees, animals. Our waste and the increasing temperatures are contributing to the collapse of our ecosystem.

- The World Economic Forum, weforum.org, reported in 2020 that one-fifth of the world's ecosystems were at risk of collapse, and most species in those areas would be wiped out.
  - This loss will cause a $42 trillion decrease in the world's gross domestic product (GDP).

Ecosystem collapse means a reduction of space on Earth to support life. It means less area for plants and animals to grow and flourish. Key indications of deterioration in our environment are already being felt and seen.

- Drought and contamination of our water supply from fires, flooding, and chemicals are so bad it's expected that by 2025 two billion people globally won't have enough water to survive.

- Wildfires have been getting longer and stronger as our world gets dryer and storms become more intense.
  - Research shows that one-fourth of our carbon emissions come from forest fires.
  - Trees naturally sequester carbon, yet some are burned or cut down to clear land for beef and our habits of convenience.
- Droughts have been increasing for years, and we're now feeling the effects.
  - Food shortages will continue to increase as we have less land and water to grow it.
- Melting icecaps are quickly raising our sea levels.
  - Flooding in coastal areas negatively impacts the tourist industry, half of which revolves around coastal communities.
- Warming weather increases storms and the risk of flash flooding, which has become more common with severe weather.
  - We are experiencing more above-average hurricane seasons.
  - A dryer climate means more erosion that contributes to flooding.
- Heat waves have gotten worse and are getting longer.
  - According to the US EPA, the average length of a heat wave lasted in the US nearly tripled from 1960 to 2020. Increasing by almost 50 days.

If you look around, everything is either really dry or super wet. So our environment is off balance.

Whales and bees - we won't survive without them

Though nature is resilient and has been withstanding human impact for generations, there is a point where it can no longer take the stress. This is called a tipping point, where we lose so much of a species that it cannot recover, and life on the planet is permanently impacted. Losing our coral reefs is a good example. Here are a few more that might surprise you.

Two of the most threatened species are bees and whales.

- Bees are the planet's top pollinator, with around 90% of most crops relying on bees to help them grow.
  - We need bees for honey, a healthy food with medicinal uses.
- Whales are one of the smartest mammals on the planet. They teach us about ocean life and how it balances all life on Earth.
  - They are important to maintaining a stable food chain in the oceans and preventing the overpopulation of certain species, like krill, which they consume in large amounts.
  - Whales contribute to the carbon-capturing abilities of our oceans.

In the last few tips, I gave ideas on how we can help save our oceans which will help save whales, but what can we do for the bees?

- Pesticides are the biggest threat to bees.
  - One way to save bees is to eliminate chemicals on plants. We do this when we buy organic produce and grow our own food.
- Another way to help save bees is with eco-friendly landscaping.

Tip 114 - Eco-friendly planting and landscaping help bees and the environment.

When you have an eco-friendly landscape, you grow native plants that work with the soil for the healthiest growth. Healthy plants help bees and insects thrive and increase nature's carbon-capturing powers.

- There are many cool designs for eco-friendly landscaping. Look for one that suits you.

- From rock gardens and ponds to green walls covered with native plants, there are a lot of options to choose from and tons of tips online to help you find what works best in your area and climate.
- Plant ground cover instead of grass. Things like moss, vine plants, or succulents don't require a lot of maintenance, can look very appealing, and reduce GHGs. You don't have to mow, and you still have a nice-looking yard.
  - Grass is not the healthiest thing to grow in your yard.
    - It requires mowing, which can be a high-emissions activity.
    - It doesn't hold moisture well when cut short, which means more watering.
- Install an eco-friendly irrigation system that will reduce water waste and help prevent erosion.
- Use leaves and trimmings as ground cover around plants. As this natural blanket decomposes, it provides nutrients and organisms that plants need.
  - Having mulch or ground cover around your plants helps keep CO_2 in the ground.

Plants absorb CO_2, so the more we let them naturally grow, the more we improve air quality and fight climate change.

- Your plants do more than provide a resting place for bees and birds; they also help clean the air.

Can Technology Save Us? That's a Big Maybe

Technology is great. It helps us understand what's happening to our environment, and it can help us fix some of the problems. But it can only go so far without our support.

- Our garbage is growing too fast. If we don't stop filling the earth with trash, no technology will be able to clean it up.

- In 2020, it was estimated that we had over 5,000 tons of plastics in our oceans. It's going to take a lot to clean this up; we can't afford to add to it.

Companies are doing great things with technology, like making sunglasses and shoes out of ocean plastics, but it's not going to take care of the source of the problem - us.

- We need to change how we do things; it's time to eliminate disposables.
- We need to show others how important change is by setting a good example and talking about waste reduction.
- We need to use our voices and wallets to make companies and governments improve infrastructure with anti-pollution, reuse, and waste reduction policies.
  - Policies and the financial support that goes with them help fund research into planet-saving technologies.

Advances in planet-saving science and technology

There are lots of cool things happening to save our planet with technology, from converting to renewable energy sources and prevention of harmful emissions to giving us ways to eliminate disposables.

- Some Materials Recovery Facilities (MRFs) are now using robots to sort recycling, which has proven to be a dangerous job for humans.
- As we build more electric cars, technology is being developed to recycle the batteries so the materials don't end up polluting our planet.
- Cities like San Francisco are leading the charge to reduce waste with zero-waste venues and pilot programs.
  - In the San Francisco Bay Area, a pilot reuse program by Upstream Solutions, upstreamsolutions.org, is using tech to microchip reusable take-out food containers.

- Through a special pick-up program, the containers are collected at residents, washed, and returned to restaurants for reuse.
- Natural gas leaks, especially from fracking, increase damaging methane in our atmosphere, that's preventable.
  - Existing leak detection technology can reduce methane emissions significantly.
    - We need environmental protection laws that require fossil fuel companies to use this technology.
- Carbon sequestration technology is being developed that can help. There are three types.
  - Biological sequestration is naturally occurring and happens when life is thriving on the planet. To accomplish this, we need to...
    - Plant trees and save the ones we have.
      - How we eat and what we buy make a big difference.
    - Work on making our soil healthy again.
      - How we grow food needs to change.
    - Have healthy oceans and sea life.
      - Burning fossil fuels increases ocean acidity and temperatures, killing sea life, including coral.
  - Technological sequestration means pulling CO_2 from the atmosphere.
    - Captured CO_2 can be used to make graphene, a material needed in manufacturing smartphone screens.
    - Scientists are continually looking for more innovative ways to capture CO_2 and use it as a resource.
  - Geological capturing of CO_2 grabs it as it's emitted from industrial sites and stores it for future use.
    - Fossil fuel companies like this because it would commit us to a long-term use of their products.
    - They should be preventing these CO_2 leaks, not using them to perpetuate our dependence on fossil fuels.

We have the technology to energize two-thirds of the world with renewables but getting us there is a big challenge.

- Big money is poured into preventing our transition to renewables by the people who make that money from fossil fuels.

Tip 115 - Tap into renewable energy whenever possible to support our transition.

People need to take a serious look at how and where they buy energy to reduce their reliance on fossil fuels. There's more to do than driving an electric car.

- Currently, wind and solar provide an estimated 10% of our energy, but they have the potential to do a lot more.
    - Renewables are more affordable, and we need to tap into them. For example, according to energy.gov, the average cost to install solar panels has dropped by 70% over the past decade.
    - Options are available to buy or lease solar panels. Do some research to find out which is the right choice for you.
        - In the US, check out the *Homeowner's Guide to Going Solar* from the Department of Energy. They give you ways to find incentives and rebates.
        - Keep eyes and ears open for opportunities. I heard an ad recently where a solar company was offering free air conditioning units to homeowners who bought solar panels from them.
- Energy companies are investing in renewables like wind energy.
    - Check with your provider to sign up for these programs.

Global efforts that are making a difference

It is exciting and hopeful to read about all the things being done to reduce waste and emissions around the globe.

- In Australia, a scientist developed a way to make building materials from clothing textiles mixed with glass.
- A French startup found a way to use enzymes to break down plastics, so they're strong enough to be used repeatedly. This is what is needed for us to be able to recycle plastics.
  - Plastics lose their strength when recycled, so currently, they are downcycled into an item that can't be recycled.
- Globally, companies have been burning trash for decades to get rid of it. Eventually, they began capturing the heat for energy use. But incineration is very damaging to our environment and our health.
  - Waste to energy (WTE) technology is a way to turn landfills into energy.
  - An even better technology is WTE gasification which turns trash into a gas that can be used for many things, including energy.
    - This eco-friendly way of developing synthesis gas, or syngas, can be the solution to two of our biggest problems - excess waste and our energy demand.
    - Unfortunately, WTE syngas can't compete with the price of fossil fuels, so this technology hasn't taken off the way we need it to.
    - Experts say whether WTE solutions grow depends on climate change policies and initiatives to help fund it.

These are a few of the ways technology is being used to fight climate change. You may know of others, or you may want to learn more about what is being developed.

- The NRDC.org newsletter, *Top of Mind,* offers an in-depth understanding of new technologies that could save our planet.

None of it works without our support

I don't know about you, but I'm not ready to leave the health and well-being of my family in the hands of people who I *hope* will figure this mess out.

Tip 116 - Look for opportunities to work for a sustainable company.

The world needs more people working on sustainability, but that doesn't mean you should ditch a job you love. On the contrary, you're healthier when you love the work you do.

- Need a career change or trying to choose your career path? Education programs and jobs in sustainability are everywhere.
    - Universities now offer degrees in sustainability.
    - You'll feel better about going to work if you're job helps save the planet.
        - Use your skills to make a cleaner, more sustainable life for future generations. We all have something to contribute.
- Love where you work and don't want to change jobs?
    - Ask questions and find out how you can help your company and coworkers live sustainably.
        - Maybe your company needs waste reduction and recycling education programs.
    - Most people are interested in helping the planet, but they don't know how.
        - Use this book as a tool to help educate others.

Let's Clean Up Our Mess Before It's Too Late

Melting ice caps and rising seas are the result of greenhouse gases (GHG) in our atmosphere, which cause increased temperatures.

Being aware of these gases, where they come from, and how we can reduce them is the responsible thing to do.

Tip 117 - HFCs are dangerous greenhouse gases that are totally avoidable.

Hydrofluorocarbons (HFCs) are GHG that is known to be ten times more damaging to the planet than CO_2. Yet, unknowingly, we are using them every day.

- HFCs have been used since the 1970s for refrigeration and air conditioning.
    - They're now illegal in many countries.
- They contribute to warming more than any other emissions.
    - Studies show that eliminating HFCs could avoid up to a half-degree of warming. That's huge!

In 2016, 170 nations signed a pact to start phasing out HFCs by 2019. In the US, the AIM Act - American Innovation and Manufacturing Act - is a bill for phasing out 85% of the import and production of HFCs by 2035.

- Several countries have already banned the use of HFCs. In 2021, the US President proposed a plan to phase out production.
- According to HFCBANS.com, 12 states in the US now have bans against HFCs, and six are proposing them.

Alternatives to HFC coolants have been available for many years but can be expensive. Legislation and tax incentives are needed to force a shift in refrigerants and coolants to non-HFCs where they haven't yet been banned.

Refrigeration and cooling leaks are a big problem. Even if HFCs are banned, it will take time to phase them out. In the meantime, something needs to be done about leaks.

- Leak prevention in refrigeration should be mandated by law.

- o Open coolers in groceries should be eliminated, and seals on refrigerator doors should be regularly maintained to prevent leaks.
  - o When refrigeration is left open, it causes units to run more, increasing HFC emissions.
- We can choose to shop at grocers taking responsibility to eliminate HFCs.
  - o Go to climatefriendlysupermarkets.org to search for grocers in your area that have replaced HFCs or are working to decrease these emissions.
- Always keep your home cooling system and refrigerators well maintenance to prevent leaks.
  - o Leaving the fridge door open does more than waste energy; it increases GHG emissions.
- Always dispose of household coolants responsibly. Green America offers tips on how to do that at greenamerica.org/coolant-disposal.
- If you are getting a new refrigerator or air conditioner, look for low- or zero-emission options.
  - o In some areas, these are now required by law, and there could be rebates or incentives for buying them.
- New construction and replacement equipment should always be upgraded to non-HFC coolants.
  - o Some building codes now require it.

The North American Sustainable Refrigeration Council, nasrc.org, is a great resource for learning more about how we can support the reduction of HFCs.

Good news - it's not too late if we change our habits today

Scientists have made it clear. It's not too late to avoid major devastation to our ecosystem, but we are running out of time.

Two myths about the climate crisis: "It won't affect me," and "There's nothing I can do about it."

Thankfully you bought this book, and so you believe there is something we can do. It starts with changing our habits and how we do things.

Tip 118 - How you bank, order products, or use the internet can be eco-friendly.

Where you bank makes a difference to the planet, your bank uses your funds to invest them, that's how they make money, and most banks invest in fossil fuels.

- It may be a hassle to change banks, but as we all shift our money to responsible companies, traditional banks will see they need to stop investing in an industry that is harming our planet and our lives.
- There are banks that are certified sustainable. This means that it's verified they don't invest in fossil fuels.

Changing how we bank and do everyday things is a great way to make sure our time and money are working for the planet, not against it. Here are a few ways you can revamp your activities to be eco-friendlier.

- Use Ecosia to search the internet. It's a search engine that plants trees when you use it and works as well as the others. Make it your default browser.
- Look for certified eco-friendly companies to do business with, like rideshare or delivery services that buy offsets to make up for the emissions they generate.
- Find an environmental non-profit that is signed up for Amazon Smile donations at smile.amazon.com. The program donates money to a non-profit of your choice whenever you place a qualified Amazon order.

Time to get social - spread the word and make a global impact

I've had people ask me, "Isn't global warming natural and the climate always changing?" A common response is that humans have accelerated climate change. But in truth, our actions have disrupted the natural balance of our climate.

While it is true that the climate is always changing, research shows that natural changes would give us a cooling planet, not a warming one. When I heard a climate scientist explain this, I realized how much our actions had caused devastation to normal weather patterns. But explaining this to other people can be hard.

It's hard to talk about climate change; even when we know we should.

When we learn how to change our habits and why it's important, we need to help others learn, too. Misunderstandings and misconceptions about what's happening to our environment can be our biggest enemy.

Tip 119 - It's time to get social for the planet. Go online and get out in your community.

We need to do more than share posts about climate change. It's time to get out and do our part to clean our planet and help others learn why it's important.

- Look for climate action organizations to get involved virtually or in person.
  - 350.org is building global grassroots efforts to reduce the use of fossil fuels and get us to 100% renewables.
    - Join the Global Climate Strike and march to save our planet. They have resources to find the next march in your area.
      - Hosted by 350.org, this effort was started by youth activists supporting Greta Thunberg.

- o Climate Reality Project, climaterealityproject.org, has local chapters building grassroots efforts within communities around the globe.
  - o National Resources Defense Council, NRDC.org, is fighting for clean air, clean water, and healthy communities.
  - o Sierra Club, sierraclub.org, is fighting to save our wild places and promotes clean energy.
  - o The World Wildlife Fund (WWF) at worldwildlife.org is one of the oldest non-profits fighting to save wilderness areas and endangered species.
- Get online and take action in your community. Find groups you want to learn from and support. Try out a few until you find one you like.

Here are a few places to learn more about climate change and talk about our impact on the earth.

- Social Media
- Neighborhood get-togethers
- Circle of friends
- Community events
- Places of worship

Use your social media account for good. Take pictures when you're volunteering to clean up the environment and post them. You may get other people interested in joining the fight.

I donate blood regularly and started posting a pic when I do. I got a thank you reply recently from an old friend who said my post reminded her to schedule an appointment. Yay!

- Wouldn't it feel great if a friend said they volunteered to fight climate change because they saw you doing it?

Most people believe climate change is real, but they don't know what they can do about it.

- By knowing the facts, we can help people understand what to do and why.

o   When we learn the facts, we know how to respond when we hear misinformation or if someone is curious about helping the environment.

Be curious. Instead of telling people they need to reduce waste or recycle correctly, ask if they know how. You might say something like, "Hey, neighbor. I'm curious. Have you checked with our recycling service to see if they take this kind of plastic?" Maybe they know something you don't, or maybe it gets them to visit your MRF's website.

Here are some key points from this chapter that will help you keep on track to fight climate change and help others do the same.

1. How we eat, buy, and grow food makes a big difference.
   a. Rethink your eating habits; reduce or avoid eating beef and fish.
   b. Support locally grown organic farming and regenerative agriculture.
   c. Growing your own food helps fight climate change.
2. Our survival depends on certain animals, like bees and whales.
   a. We need to clean up oceans and fight agricultural practices that pollute the planet and kill bees with pesticides and other chemicals.
3. Energy waste and emissions are no joke. Every bit we conserve reduces the damage caused by our increasing demand.
   a. Watch out for things that suck energy, like vampire devices.
   b. Stop using gas appliances and grills that cause harmful emissions.
4. Advances in technology can help, but only if we reduce consumerism and waste.
5. Jobs and business opportunities are growing in the green economy.
   a. How we browse the internet and bank should be eco-friendly.

Chapter 6: Build A Lifestyle of Caring for The Earth and Each Other

Forced change is never fun or easy. It is inconvenient and stressful not to be able to maintain our own personal status quo. But the cascading effects of the human impact on our ecosystem are forcing change.

- Fire seasons are no longer *seasons*; they are all year round.
- Floods from rising seas and extreme weather are killing people around the globe.
- Recovering from storms, fires, and floods costs billions in taxpayer dollars and personal loss.
  - Devastating weather disasters have more than doubled in the past five years compared to the previous three decades.

There were 432 disastrous natural events in 2021 recorded by the Emergency Event Database (EM-DAT). These caused 10,492 deaths and affected over 100 million people at the cost of around 252 billion US dollars.

Let's Step Up and Make a Difference for Tomorrow

We're past the point where we can afford to be passive about our environment. If all we do is make a conscious effort to reduce the amount of trash we generate, that can make a huge difference.

Deciding not to act simply because it's inconvenient won't work anymore.

Taking steps to build a just and better world

The opportunity to live a clean and healthy life is important to healing our planet and humanity. But that has been stripped from many people in areas where corporations build polluting plants and dump chemicals.

The fact is the least healthy areas to live in are the cheapest. I read stories all the time about people suffering because of where they live. They didn't choose to have a plastics plant or a landfill in their backyard. They live where they can afford to—people who can afford to move out of these areas.

- There's constantly new research about the increased rates of childhood asthma and birth defects tied to the environment.
  - Research has shown that California wildfires increase health risks to pregnant mothers and their unborn babies by as much as 20%.
- Disadvantaged communities are the most impacted by consumerism and our waste habits.
  - They are often near landfills that emit methane gas and incinerators that burn plastics.
    - This causes increased health problems and a rise in healthcare costs for everyone.

Reducing our waste and energy use can reduce health risks, especially for children whose bodies are still developing and are more likely to get sick from environmental factors.

- If we use less energy, power companies can shut down dirty coal and gas-burning power plants and increase the use of renewable energy sources.
- If we produce less garbage and food waste, there will be fewer toxins in the air from burning trash and methane from landfills.
- If we use fewer plastics, we can significantly reduce CO_2 emissions from making and disposing of them.

Tip 120 - Learn about climate justice and help those who live in unhealthy areas.

Everyone should have access to nutritious food, safe drinking water, and clean air. But today's reality is you can only obtain these things if you can afford to live where they're available.

People who live in disadvantaged communities should not be alone in their fight for healthy living conditions. They should not lack resources and opportunities to live a better life.

We all need to stay informed and join the fight.

It's time to step up and vote for the environment. It's time to support a better life for everyone. Find ways to get involved in helping those on the front lines of this climate crisis. If you are one of those people living in a toxic area, there are many groups that can help you fight for your rights.

- **Climate Justice Alliance**, climatejusticealliance.org, is helping threatened people around the globe to transform their communities and fight economic and climate injustices.
- **Zero Hour**, thisiszerohour.org, is a youth-led organization fighting for climate justice. Learn about and support their fight. Follow them on social media.
- **Earth Guardians**, earthguardians.org, trains young people around the world to lead climate and social justice movements.
- **MIT's Climate Portal** has a section dedicated to keeping us informed about climate justice. Go to climate.mit.edu/explainers/climate-justice.
- Learn about the six pillars of climate justice from the **University of California**, centerclimatejustice.universityofcalifornia.edu/what-is-climate-justice.

Avoid the spiral of silence - talking climate isn't taboo

A group of climate experts was asked what individuals can do to help reverse global warming. The two most common answers were about climate change and taking action in our homes and communities.

The earth needs you to speak up.

Tip 121 - To fight climate change, we need to learn how to talk about climate change.

We need to speak up for the planet. This can be hard when social media is full of judgment and misinformation. Remind yourself that the planet needs your voice.

- Most people want to know how to fight climate change; ignore the ones that don't and move on.
    - You'd be surprised how many people care about saving the planet.
- Keep it simple. Avoid over-explaining why you're changing your habits.
    - If people ask, say, "It makes sense for the planet." Some people may want to learn more and ask questions. This is how we make change happen.

Social media is tricky because most of what is shared is opinion and incorrect information.

Don't share a post unless you verify what it says is true

The best way to combat misinformation is to learn and share the facts. Here are a few hashtags you could use to promote a healthier planet: #globalwarming, #reducewaste, #recycling, #recycleright, #rethinkrecycling, #zerowaste, #climatechange, #climatejustice.

When the Impossible Becomes the Inevitable

It seems impossible that life as we know it could disappear. But with occurrences like floods, severe storms, fires, and sudden extinctions, it's beginning to feel possible. Not in a few hundred years, but in a few decades.

- On the flip side, I'm a *glass-half-full* kind of person. Humans have always been up for a challenge. The climate crisis will be our greatest yet.

The climate crisis has increased the demand for alternative energy sources. While this transition should have been started a few decades ago, more is happening now that the need has become obvious.

- Renewable energy now provides over a quarter of global power.
    o This isn't growing fast enough to meet rising demand.
- Energy demand continues to rise. Everyone, from big companies down to individuals, must reduce energy demand and push toward renewables.
    o Eliminating energy waste will help.

Addressing waste is one way to reduce our energy demand; another is to compensate for it. This is often done through offsets, though they're not always effective.

Tip 122 - How carbon offsets work and when they may not be effective.

The more aware we are of our climate crisis, the more companies are scrambling to look sustainable. They don't want us to stop buying from them. Some purchase carbon offsets to make up for their emissions from manufacturing and shipping. These are great if they work.

Over 80% of us want to buy from companies that are sustainable.

Buying carbon offsets doesn't make a company sustainable unless those offsets are effective.

- Tree planting is a common way to offset CO_2 emissions, but nothing is gained if the trees aren't protected from logging or forest fires.
    - Trees need to live long and prosper if they're going to heal the earth as we need them to.
- If a company invests in carbon reduction projects that would have happened anyway, it doesn't qualify as an offset.
    - We need more carbon reduction projects, not just support for existing ones.
- Are the offsets supporting current or future projects that are speculative and not guaranteed to reduce emissions?
    - This is called hedging and doesn't offset current emissions, which is what we need.
- Do they prevent emissions or only shift them to other activities that still cause greenhouse gases (GHGs)?

If a company says they buy offsets, it should detail the type and source, especially if they're a publicly traded company. Some companies buy offsets to look good to investors and customers; they may not care if what they buy is doing the job.

The more we buy from sustainable companies, the quicker we change the one-way stream of trash into the earth and damaging emissions going into the air. So let's buy from companies that verify they're sustainable and aren't greenwashing.

Tip 123 - How to tell if a company is sustainable or are they greenwashing.

It's hard to believe what companies say because they only tell us what makes them look good. Sometimes the truth hurts. We must see through their marketing and do the research if we want to give our money to environmentally responsible manufacturers.

Some companies say they're sustainable when they aren't.

Companies can make marketing claims and tell us what we want to hear, but these don't mean their operations are sustainable. They need to back those claims with evidence.

When a company makes it look like they're sustainable with buzzwords and unsupported claims, they are greenwashing.

Greenwashing uses marketing and press releases to persuade consumers that an organization's products and policies are environmentally friendly.

Here are things to look for when you want to verify if a company or its products are sustainable.

- Where do they get materials, are they recycled or new (virgin) materials?
- Do they use organics, like cotton, for clothing and avoid synthetics made from fossil fuels?
- Are their facilities energy efficient and using renewable energy?
- Do they minimize waste of energy, water, and materials in manufacturing?
    - Are their supply chain factories zero waste and zero emissions?
- Where do they make their products?
    - How much shipping is involved?
    - How fair are their labor practices? Do they pay a living wage?
- If they say they buy carbon offsets, do they provide proof that they work?
- Read their sustainability page and look for the important information that verifies their claims.
    - Do they belong to organizations that confirm they're sustainable?
    - Do they address the impact of their manufacturing, like CO_2 emissions or chemical and water use? Can they prove it?

If a company is getting certified for its practices, they're likely serious about sustainability, but they need our help to keep it going. We need to support them by buying their products.

There is no silver bullet that's going to fix this - it's up to us!

Yes, new technologies can help us transition to a more sustainable life, but this will only happen if we're all on board. We can't sit back and point fingers at others, saying it's their problem. This problem belongs to all of us, no matter who caused it. Moving forward and fixing it is all we should focus on.

Sustainability is a way of life we all need to embrace.

In reality, businesses build manufacturing plants and mine for materials to make money from us, the people that demand their products. So, we control what they do.

Legislation and Big Business - Stay Informed and Vote for the Planet

More tax dollars go to dealing with our trash than building infrastructure and regulations that will make recycling work. That's dumb.

To protect the planet is to protect the people that is the purpose of the government.

Pay attention to local elections if you want to make a difference in how climate-friendly your government is. For example, packaging and recycling laws usually get enacted at a local level. Better packaging laws mean better recycling and less confusion; we all want that.

Research your local politicians and send them emails telling them you want effective climate legislation in your community, like Extended Producer Responsibility (EPR) laws.

- When it's election time, I turn to established environmental groups, like the Sierra Club, to find out who they endorse.
  - There are many organizations that evaluate propositions and politicians to determine if they are good for the environment. These are great resources to make sure your vote helps the planet, not hurt it.
- We may get frustrated with politics, but that should be the reason we speak up and get involved, not turn our backs in disgust and not vote.
- Over 60% of people in the US believe that climate change is impacted by human actions.
  - Less than 20% of voters write to their government representatives urging them to support climate action.
    - The Climate Action Now app makes this easy.

Without us speaking up for the planet, big money and big business get a stronger voice, and their interests are often in profits, not the planet.

Because people are speaking up for the planet, governments are starting to change.

- Canada has banned single-use plastics.
  - In December 2022, the country will no longer allow the making or importing of disposable plastics.
  - By the end of 2023, all sales of this packaging will no longer be allowed.
- California now requires transparency about where recycling goes and how it's handled.
  - The goal is to divert 75% of waste from landfills, which can only happen if recycling is tracked and not shipped overseas.
  - The law requires truth in labeling.
    - Soon, the recycling triangle will only be allowed on materials that can be recycled.
    - No more fake recycling claims.

- Colorado's governor signed a statewide ban on plastic shopping bags and polystyrene carry-out containers that will go into effect in 2024.
- Local governments around the globe have enacted EPR laws that will require companies to take responsibility for their packaging and what happens to it.
  - Germany has had EPR laws in place since 1991.
  - In Canada, nine out of the ten provinces have EPR laws.
  - In the EU, many member states have had EPR laws for years.
    - They will be required in all EU states by 2024.
  - Two states in the US recently enacted EPR laws; Maine and Oregon.
    - These go beyond bottle deposit laws and apply to all packaging.
  - Australia first introduced container deposit schemes, a version of EPR, in 1977.

Tip 124 - Let legislators know the health of the planet needs to be their priority.

For some time, we've had the technology to eliminate gas and diesel combustion engines in public transportation and trucking. Since these are the biggest emitters of greenhouse gases, this would significantly slow down and could reverse rising temperatures.

We need policies that will force the transportation industry to convert to zero emissions. This only happens when we speak up and tell our elected officials to act on behalf of the planet and the people.

- In the US, find your government representatives by visiting www.usa.gov/elected-officials
- In the EU, the European Environmental Agency, eea.europa.eu, provides information on the latest environmental initiatives in European countries.
- Investigate your local politicians and candidates to find the ones who support climate initiatives.

- o Go to vote-climate.org to check the climate position of political parties in the US.
- Check out the Sierra Club list of climate champions at sierraclub.org.

There's big money being spent by the fossil fuel industry to prevent the growth of renewable energy and promote the increased use of plastics. Our planet won't survive if they continue to get away with it.

- In the US, 16 states have laws to prevent bans on plastics. These happened because fossil fuel lobbyists pushed politicians to pass them.
- Some governments pay more money to businesses that ship recycling overseas rather than manage it at home, at the source.
  - o These subsidies do not take care of the problem. They only make it worse.

Groups like The Institute for Local Self-Reliance, ILRS.org, provide tools for municipalities to build strong circular economies around recycling that create jobs.

- Visit the Waste-to-Wealth section on their website to learn more.

The Green Economy - A Future Full of Possibilities

There are those who believe that changing from a disposable lifestyle and dependency on fossil fuels would mean economic hardship. Nothing could be further from the truth.

There is more support than ever from governments as countries step up to curb emissions over the next two decades. This means investments in infrastructure.

- Building electric vehicle charging stations along highways.

- o   This will support a transition to electrified trucking.
- o   Faster charging technology will make this more practical.
- Better recycling facilities and more of them.
  - o   A shift from fossil fuels means tons of jobs to develop and provide effective recycling of plastics and other materials so we can start reusing them.
  - o   Investment in universal programs that provide consistency in recycling will eliminate confusion and make it easier for us to do it right.
- New technologies are emerging to tackle this large clean-up project we have on our hands.
  - o   Not only the trash that is choking our planet but also cleaning up the way we do things.
    - ▪   We will see investments in alternative ways of doing things, which will eliminate emissions and chemical contamination.
  - o   Facilities and infrastructure need to be built, and people are needed to design, engineer, construct, and run them.

Building the infrastructure to support a sustainable planet will create jobs in wind and solar power, engineering, construction, automotive, and biofuels, to name a few.

- The United Nations projects that the green economy will create 24 million new jobs.
- Jobs are being created as we build renewable energy. The goal is 80% clean energy by 2030, more than twice what we currently generate.
  - o   The good news is we have what it takes to do this, and it requires skilled workers.
    - ▪   Look into education grants that will pay for you to become one of these in-demand workers.
- Over the past decade, US jobs related to the solar industry have increased by nearly 170%.

From clean-up to new technology, jobs will be abundant as we move into the green economy.

Tip 125 - Opportunities for entrepreneurs in the green economy.

Opportunities around the world are exploding as the demand grows to reduce waste and do sensible things with the materials we have. So if you're looking for a good business opportunity, saving the planet has a lot to offer.

- Eco-friendly planning and landscaping are in demand.
    - More people want ground cover and wild yards instead of lawns.
- Organic farming is on the rise.
    - As governments work on fixing our agricultural problems, tax dollars will need to be invested in organic farming.
- Design and engineering firms with expertise in LEED buildings (Leadership in Energy and Environmental Design) are needed.
    - This green building rating system is globally recognized.
    - There are often subsidies available for green construction projects.
- Recycled construction materials are a growing business.
    - I read about a company that buys recycled plastic and turns it into decking material. Plastic is strong and weather resistant. Great for building a deck.
- Textiles and furniture create unique opportunities for creatives. It can be an inexpensive way to open a recycling business.
    - Think Etsy. People visiting this site want unique handmade items.
- E-waste recycling can be very profitable.
    - Careful handling is required, and training is needed, but the recovered materials, like cadmium and phosphorus, are quite valuable.
- Metal is one of the oldest types of recycling that is still in huge demand.
- Companies like Refrigerant Finders, refrigerantfinders.com, buy used and surplus refrigerants for repurposing or destruction and have made it profitable.

Corporate responsibility and the leverage we have over it

To live sustainably, we must invest in companies that act responsibly and work to be sustainable. When we give a company our money, we need to make sure they aren't using it to harm us.

The fact that 100 companies are responsible for over 70% of emissions is unacceptable.

Stockholders of publicly traded companies have a say in how they do business.

Tip 126 - Take a sustainability health check on your investments.

If you own stock, you have voting rights you can use to encourage the company to lower its environmental impact. This strategy is making headway in corporations where shareholders are speaking up and using their votes for the planet.

Make sure your money isn't being invested in fossil fuels.

Investing in the fossil-fuel industry means supporting it and the damage it causes. There are many ways to de-fossilize your investments.

- Most investment firms have options to switch your money to sustainable and responsible companies.
  - These investments include your 401k or IRA retirement funds.
- Green Century Funds offers investments in sustainable companies.
- Search for sustainable Muni ETFs (exchange-traded funds) that put your money in tax-free municipal investments that support sustainable development, as well as social and environmental change.

- Look at the environmental, social, and governance (ESG) policies of the companies you invest in.
    - You want to make sure what a company says is true. Investigate them and make sure they aren't greenwashing. **See Tip 123**.
    - CSRHub.com has a database that rates corporate ESG practices.
- Search for eco-friendly investments that are profitable. The green economy is a booming economy. There are many places you can invest where your money can work for the environment, not against it.

CSR = Corporate Social Responsibility

ESR = Environmental, Social, and Governance

Sustainability is a smart investment, and you'll feel better about the money you're putting away for the future when that means a better future for all of us.

The Good That Comes When We Think Before We Buy

How human activity impacts the planet and climate change comes down to two things.

1) How companies make products and the waste caused by that production and,
2) How products are consumed and what happens to post-consumer waste.

This applies to everything we buy, from products and food to energy and fuel. It even applies to products we don't want and return.

Perfectly good materials are going to waste.

When you return a product that isn't defective, you expect it will be resold, but this is not the case for nearly half the products returned.

- Because of the hassle of sorting out the good returns from the defective ones, many retailers send returns to landfills rather than back to store shelves.
- This is a good reason to think before you buy.
  - Even if they offer free returns, the damage caused to the environment means it really wasn't free.

Tip 127 - Buy from zero-waste and eco-friendly retailers whenever you can.

Grocery and home goods delivery has become very popular because it's convenient. But, if you use these services, find ones that are sustainable with waste reduction and effective carbon offsets.

Several online retailers offer sustainable products and packaging. We need to buy from them to keep these businesses growing and available.

- Grove Collaborative, grove.co, is an online retailer of natural personal care and household items. They are plastic neutral and target having plastic-free packaging by 2025.
- Hive Brands, hivebrands.com, sells sustainable groceries and household items. They evaluate each product they sell for its ingredients and packaging. In addition, their shipping is carbon neutral, which means they use effective offsets.
- Patagonia, patagonia.com, offers more than environmentally conscious sporting goods. They now have Patagonia Provisions, a line of eco-friendly food products for outdoor adventures and camping.
- Preserve, preserve.eco, makes self-care products from recycled materials, as well as food storage and kitchen products. They're climate neutral certified.
- Seventh Generation, seventhgeneration.com, offers plant-based products like cleaning and household goods that can be found online or in stores.

Tip 128 - Reduce the impact of chemicals, and skip bright colors in clothing and paper products.

- We don't always think about the impact of chemicals used to make our products. But where these chemicals end up has a significant impact on our lives.
  - Chemicals contribute 20% of the greenhouse gases from landfills.
- I used to buy paper towels with a pretty print; it livened up the kitchen. I didn't think about how the chemicals used to make them poison the earth.
  - Since paper towels are not recyclable and have to go in the trash, the dyes used to color them are harming our environment.
  - When I found this out, I thought, *now that's something people can control.* We can choose to use cloth napkins instead of paper or buy white paper towels and napkins without color print.
- Bright-colored paper can't be recycled. Only office paper and lighter colors can. The rest end up in landfills.
- There's a reason why clothes that are made sustainably are often neutral colors.
  - Colorful clothing might be fun to wear, but it's not fun for the planet. If they're not all-natural, dyes used in clothing are bad for the planet and contribute to landfill emissions.
    - Look for products with the bluesign® label, which means they're monitored to make sure they don't use harmful chemicals. **See Tip 130**.

Tip 129 – Find out how green your favorite brands are.

It feels good to buy things that will cause minimal impact. A way to do that is to know how the product was made and how sustainable the retailer is before you buy it or buy it again.

- Ask your search engine, hopefully, Ecosia, how sustainable your favorite brand or retailer is. Look for articles and reviews on their environmental practices.

- Find the *Sustainability* page on the company's website. If they don't have one, that tells you something.
  - Look for their CSR or ESG report for data on what they do to...
    - reduce emissions,
    - reduce waste of materials, water, and energy,
    - eliminate the use of harmful chemicals and...
    - responsibly source organic and recycled materials.

They should answer these questions for their company-owned operations and their supply chain. Since the making of products and the materials used is one of our most damaging activities, a company's supply chain needs to be transparent.

- There are third-party organizations that monitor manufacturing practices. From the materials used to how responsibly they use resources and reduce waste.

Know what's in everything you buy.

Companies need to be responsible for the entire life cycle of their products, from the time the materials are made to what happens to products when they're no longer in use.

- To determine if a product is eco-friendly, we need to know its impact from the cradle to the grave. This is called a lifecycle analysis.

Here are resources that can help you determine the impact of the products you buy. If a company doesn't show up when you search these sights, they aren't likely doing what it takes to prove they're sustainable.

- **B Lab** is a non-profit that sets standards and certifies corporations that act responsibly toward society and the environment. Visit bcorporation.net and go to *Find a B Corp* in their menu.
- **FairTrade.net** monitors labor practices in manufacturing facilities and supply chains to verify workers are treated

fairly and receive a living wage. This is important to living sustainably.

- **OnePercentForThePlanet.org** will tell you if your favorite brand is among the over five thousand companies that contribute 1% of sales to helping the environment.
- **SustainableBrands.com** offers the latest news and reviews on what companies are doing to be more sustainable.
- **SustainableJungle.com** lists ethical and sustainable online stores that provide home goods, beauty supplies, clothing, and groceries.
- **TheGoodTrade.com** provides reviews and resources on home decor, fashion, and sustainable living.
- **EcoWatch.com** reviews products to verify how sustainable they are and provides industry news updates.

Look under the name of the parent company if they're owned by a bigger corporation. You should be able to find this on the *About* page of a company's website.

Our dollars speak volumes. It's why companies are in business. When we buy from sustainable companies, we tell them we support what they're doing and we support a healthier planet.

Tip 130 - Avoid fast fashion and disposable clothing trends.

The apparel industry is our third largest global polluter, thanks to fast fashion and social media. From manufacturing to a product's end of life, when it usually goes to a landfill, there are big environmental impacts caused by the clothes we buy.

- The garment industry alone is responsible for 10% of our global emissions, second only to the oil and gas industry.
- More than 80% of the clothing we buy ends up in landfills.

Here are some resources to help you buy apparel made from responsibly sourced materials and manufactured sustainably.

- Bluesign® verifies that apparel companies follow best practices for chemical, water, and energy use. If a brand carries the bluesign® label, they meet these standards.
  - If your clothes don't carry the label, check bluesign.com to find out if your favorite brand is in the process of being certified as a systems partner.
- **Cosh.eco** is a European site that analyzes fashion brands for their sustainable practices.
- **Global Organic Textile Standard (GOTS)**, global-standard.org, sets organic textile standards and certifies apparel companies that meet these.
- **GoodOnYou.eco** reviews thousands of fashion brands to tell us which ones are the most sustainable.
- **OEKO-TEX.com Buying Guide** verifies apparel companies for their sustainable textile and materials usage.
- **Worldwide Responsible Accredited Production (WRAP), wrapcompliance.org**, certifies factories in the sewn products supply base on their social compliance and provides a search map to find these suppliers.

Join the Fashion Revolution and ask your clothing manufacturers on social media: #whomademyclothes.

If you like writing reviews about companies and products, make sure to include positive comments about the sustainable companies you buy from and why you choose them.

- An example might be, *I only buy XX brand because I know they contribute one percent of all sales to saving the planet.*

Changing how we eat can heal our ecosystem and our lives

Excess consumption of meats like beef and fish is damaging our planet. When we reach a point where we burn down trees, pollute our oceans, and basically throw off our entire ecosystem to meet the demand for certain foods, we must rethink how we eat.

- One study showed that 57% of food's carbon footprint comes from meat and fish.

How and what we eat makes a big difference to our environment.

Tip 131 - Eating fish and shellfish causes more damage than you may realize.

The biggest threat to ocean life is human activity. The worst is from fishing practices. Though studies may say that eating fish is good for you, the way we catch it isn't.

- Bycatch is unwanted sea life caught in fishing nets. These animals are not rescued; they are slaughtered because it is inconvenient to save them.
  - Marine mammals that live near the surface to get air, like dolphins, are most at risk of dying as bycatch.
  - Studies show that fishing practices are the deadliest of all human activity to ocean mammals.
- Seafood brands that say their fish is sustainably caught may not be telling the truth.
  - These claims aren't monitored, even though there are regulations about their use.

Overfishing and the industrialization of the fishing industry have caused a major scar on our oceans that is turning into a deep, gaping wound that we may not be able to heal.

90% of ocean mammals have been wiped out.

- Do research before trusting a label that says sustainably caught.
- Consider cutting back or completely cutting fish out of your diet.
- A recent study showed that one in three fish coming into the US was illegally caught. Meaning other sea life was killed in the process.
- Sea Food Watch, seafoodwatch.org, is a good resource to find out what to avoid and how to buy sustainably caught fish.

There are people trying to fix our fishing problems; they don't need us working against them. That means we need to reduce our demand for fish until our oceans and seafood are healthy again.

Eating less seafood means more life in the oceans and a better chance of healing them.

Tip 132 - How we raise and use livestock is very damaging.

Like many of us, I grew up in a home where meals consisted of meat, starch, and veggies. The *meat and potatoes* lifestyle. This mentality is a major contributor to our unhealthy planet.

Livestock farming is the biggest cause of dead zones in our oceans.

There's a reason many people are turning to a plant-based diet. The most significant is the devastation caused by livestock farming. Meat is a luxury we can no longer afford.

- Animal farming is responsible for nearly 20% of greenhouse gases which are directly related to rising ocean temperatures that cause coral bleaching.
- Livestock poop is loaded with antibiotics and hormones and often gets spilled into our oceans and waterways.

This is not the only damage our meat consumption does.

- Cows take up a lot of land.
  - 70% of the Amazon Rainforests have been cut down to make room for cattle grazing.
- The meat and dairy industry uses almost one-third of our fresh water.

Consider eating a plant-based diet or going meatless a few days a week for the sake of the planet.

- If you're not ready for a plant-based diet, why not skip meat a few days a week?
  - Try a meatless pasta dish and salad for your meals.

- If you enjoy eating meat, try cutting back on beef.
  - Eating beef only once a week will make a huge difference, especially if everyone does it.

Studies are proving that eating a plant-based diet is better for your health and the environment.

Tip 133 - When shopping for a product, compare locally made to online stores.

Once you know what ingredients or foods to avoid, you'll need to find alternatives. Sometimes the best options are from an online store; sometimes, they're not. A lot depends on where you live. Take time to compare and find your best choice to help the environment.

- There are several online stores that are zero waste and use recycled packaging. If you don't have local grocers that offer these, you might want to buy your groceries online.
  - Shipped to your door or shipped to the grocer down the street, either way, creates emissions, but an online store is more likely to buy carbon offsets.
  - **See Tip 127** for sustainable online grocers.
- Check if the foods you buy are made locally.
  - Look online for butcher, dairy, or produce farms in your area. Then look for their products in stores.
    - A local producer's website will say what stores carry their products.
    - If you buy locally made, you support your community and reduce emissions.
  - Get familiar with your local farmer's market. It's a great way to reduce the environmental impact of your food and support your local economy.
- Do a sustainability health check on the stores in your area where you shop. Here are some things to look for…
  - Do they source locally?
  - Do they carry organic and plant-based products?
  - Do they sell products with reduced and recycled packaging?
  - Do they use renewable energy?
  - Is there refrigeration well-sealed and energy-efficient?

- Do they use HFCs for coolant? These are harmful chemicals that are illegal in many countries. **See Tip 117.**
- Build a list of local and online retailers with sustainable products and packaging.
- Write the brands you've investigated on your shopping list, so you know what's safe to buy.

You will live and feel healthier - think balance and minimalism

Eating unprocessed or minimally processed foods, like locally grown fruits and vegetables or whole grains, is good for your health and the health of the planet. Unfortunately, one-third of global emissions come from our food production, and a big contributor to that is processing.

Tip 134 - A balanced diet is better for you, your family, and the environment.

You've heard it before, eating a balanced diet is good for you. But it can also be good for the planet.

While some food processing may be necessary, like pasteurizing milk, most packaged and processed foods aren't good for you. Studies show that people who eat ultra-processed foods consume more calories and gain more weight. This affects how long you live and your quality of life.

More processing means more waste, emissions, and chemicals. So, it makes sense that doing this to our food is unhealthy for the earth, and eating more whole foods is not only good for your body but all living things.

Here are ways to eat a balanced diet with minimal processing.

- Meal planning is a great way to determine what to buy and not spend extra money.

- Always include fruits and vegetables in your daily diet.
  - Local grown should be your first choice when possible.
- Buy organically grown and minimally processed plant-based foods.
- Visit Food Print, foodprint.org, to calculate the impact of the foods you eat and get tips on eating sustainably.
  - They fight for transparency in our food production so we can make informed choices that are healthier for everyone.
- Whole grains mean little processing, fewer emissions, and fewer chemicals in our food.
  - Did you know flour is not naturally white? It's bleached, which is how we get white bread.
- Cutting back on eating meat is easy with protein alternatives like soy and legumes.

The good news is if you're already eating whole and low-processed foods, you're helping reduce the impact of food production on the environment.

Tip 135 - Use plant-based products to replace harmful chemicals in your home.

Most cleaning and household products are made with chemicals that are harmful to the environment. Whether they go down your drain or end up in the earth, they're causing damage.

- From your hand soap to your window cleaner, if they aren't all-natural, they damage the environment.
- Buy plant-based products whenever you can.
- **See Tip 127** for ways to buy natural and plant-based household products.

Tip 136 - Buy organic clothes - don't wrap yourself and your family in chemicals.

I'm old enough to remember when polyester clothing first became popular. It gave us cheaper apparel that didn't need to be ironed. I don't like to iron. I never thought to wonder how polyester is made.

Then again, I also remember a lot of talk and regulations around toxic chemicals in clothing back then.

- I've never trusted the levels of toxic chemicals my government says are *okay* for me to consume or wear. It doesn't feel like *any* would be good.

Don't get me wrong. I'm not perfect and still own polyester, mostly because I buy my clothes second-hand and need to find something that fits. But it creeps me out to wear clothing made with fossil fuels. Now that I know the truth, I look for cotton and blends.

- There's a lot of value in buying 100% cotton.
    - Synthetic materials like polyester and nylon are made from fossil fuels.
    - Though there are rules about the levels of toxic chemicals that can be in the clothes, those chemicals are still present.
- When clothes are washed, fibers break down and end up in our water.
    - Microplastics are now found in our oceans and sea life. They're also now found in most life forms, including humans.

Tip 137 - Living a minimal lifestyle is doable. Start with baby steps and keep going.

It's easy to learn how to live a minimalist lifestyle by searching the internet on the topic. You'll likely run into videos or podcasts from The Minimalists, two guys who practice minimalism and provide great advice on the topic.

- The first step is to act. You're reading this book, and that's a great action. There are plenty more that can help.
    - Commit to reading at least 15 minutes a day about ways to reduce your impact on Earth.
    - Focus on the next action you're going to take. Whether it's changing how you eat or shop or a switch to showering every other day, figure out what's going to work for you.

- It can be hard to bring a new habit into your life, but like anything, a little bit of effort will go a long way.
- Don't overthink it; you're likely not to act. If you need a simple tool to get into action, turn to Mel Robbins and her 5-Second Rule.
  - The more of us who launch ourselves into action, the more likely we will win this battle.

When I began to recognize my occasional habit of emotional shopping - *we all do it, just admit it* - I realized the regret over wasting money on something I didn't need was far worse and longer felt than the temporary rush of getting something new.

- Then I just had another item I would eventually have to deal with and,
- I realized that how I eventually got rid of it would impact the planet.

I get it; buying new stuff can be fun *and* therapeutic. But did you really need it? If you're an emotional shopper, the answer is often "no." Browse a thrift store instead. It can be just as fulfilling and stress-relieving.

- If you buy something used, it won't contribute to harming the environment the way a new item would.
  - You'll spend less so you won't regret it as much.

Our individual efforts can go a long way

It is clear we cannot keep doing what we have been doing; wasting resources, burning fossil fuels; and polluting our planet. These human actions are causing catastrophic changes to our ecosystem that are making the planet uninhabitable.

- In many regions of the world, people have been forced to migrate from their homes due to drought or flooding.

Being a creature of habit can be a good thing or a bad thing. However, when it comes to the way we consume things, both food, and products, our habits are mostly bad. We've built up some bad habits, and we need to ditch them.

Disposable everything is not healthy, and it's proving deadly.

Start simple and move forward with little changes to make it easier to develop Earth-caring habits. The more we improve, the quicker we will be able to bring our climate back to normal.

1. Find what you can do to support climate justice.
    a. Part of caring for our planet is caring for each other.
2. Talking climate can't be taboo. In fact, it may be the only thing that saves us.
    a. Look online to learn how to talk about the climate and what messages we need to share.
        i. Keep it simple.
    b. Learn how to teach others to do the same.
3. Your vote matters to the planet.
    a. It's time we get our elected officials working for us, not against the planet.
    b. Research your local representatives and initiatives to make sure they benefit the environment, not harm it.
    c. Tell the companies you own stock in that you want them to act sustainably.
4. Find out how green your favorite brands are, and make sure they're acting responsibly.
5. How and what we buy makes a difference. From food and clothing to household goods and cleaning supplies.
    a. Make sure you're buying from sustainable companies and buy locally whenever possible.
    b. What you buy and how you eat can promote a better life for everyone.
    c. Avoid fast fashion. Buy long-lasting apparel made with recycled materials.
6. Living a minimalist lifestyle is the healthiest thing we can do for our planet.

Chapter 7: Educate and Activate - There's Always More to Do

Changing how we impact the planet requires knowledge and community. To maintain life as we know it, we will need to learn how to live together sustainably.

- **Sustainability means** living with balance in three key areas, called the pillars of sustainability. These are the environmental, economic, and societal dimensions of our world. No one wins when these are out of balance.
    - As our environment begins to crumble, economic and societal hardships are increasing.

The increasing number of global climate refugees should be a wake-up call to everyone living in developed nations and with the money to move from impacted areas.

- Poor farmers in South America aren't the consumers causing emissions, yet they're paying the price.
- Choosing to eat healthier, buy smarter, and create fewer emissions will make a significant difference to the people suffering from our past poor decisions.

This is the obligation of those of us who are privileged enough to be able to walk into a grocery store and buy food or a department store to buy clothes.

Tip 138 - Know what causes greenhouse gases.

To reduce greenhouse gas (GHG) emissions, we need to know where they come from. The two biggest culprits are...

- Carbon Dioxide (CO_2) from...
    - Transportation - shipping and freight are the biggest emitters.
        - We have the technology for this to be all-electric.

- Incentives are needed to make it happen.
  - Cutting down trees that naturally sequester carbon.
    - We can control this with regulations and better buying habits.
    - Cutting back on meat eating will significantly reduce deforestation.
  - Production of goods.
    - About a fourth of our CO_2 emissions are caused by making goods and the raw materials that go in them.
  - Energy use - we control how much we use and waste.
    - We have the power to reduce how much we use in our homes.
      - Participate in energy reduction programs and education. Your energy provider is a great place to find these.
    - Corporations need to be held accountable for their energy use and how much they waste.
      - We influence this with our voices and our wallets.
- Methane, which has 25 times the global warming effect as CO_2, comes from...
  - Agriculture: farming practices like turning soil and using chemicals instead of organics for pesticides and herbicides.
  - Livestock: it's not only cows burping that creates methane.
    - Organic waste, like manure, emits methane even when it's buried.
      - Manure is often buried rather than used as fertilizer as it should be.
      - Eating less meat is the best way to reduce methane emissions.
  - Landfills cause 20% of methane emissions and are the third biggest source of GHGs.
    - Food waste causes most of the methane emissions from landfills.
      - If we eliminate food waste, we will reduce GHG emissions.

- Natural gas leaks are significant and preventable.
    - The technology exists to prevent methane leaks, but the companies causing them won't spend the money unless they are forced to.
    - A good reason to pay attention to your local elections is to make sure corporate polluters are held responsible.

Governments need to better regulate the fossil fuel industry to protect the people, especially those that live near extraction facilities and processing plants, but this doesn't always happen.

- One disadvantaged community in Texas is trying to fight fracking and natural gas production in their area because it's making people sick.
    - When a fossil fuel company went to build more plants in their neighborhood, the residents took them to court to require they keep emissions down and protect their health.
    - The state's Environmental Protection Agency (EPA) declared it would be too expensive for the corporation to prevent excess emissions.
        - The EPA put a price on the lives of the people living in that area.
        - They fail to do their job and protect the people who pay them.
    - The only way to prevent this is to vote and make sure local officials do what they're hired to do, protect citizens from harm.

Control over most GHGs is in the hands of corporations, so they are the ones who can prevent them. That is why we need to tell our governments and the companies we buy from that we want them to take responsibility and make the changes needed to protect our environment. We need them to put people before profits.

Learn How to Live Sustainably and Stay on Track

We've discovered the power of the internet and social media to make a movement happen. We've used these to help people in need and to organize change. Now we're leveraging them to fight climate change.

Get involved and learn the facts - volunteering helps

Anyone can find opportunities to clean up their community, and we all have the power to push governments to set aggressive climate goals. So go online to find resources or local groups to start volunteering today to save the planet.

Join a movement and fight for Earth. It's the most important cause you'll ever support.

Tip 139 - Volunteering is a great way to stay current on environmental issues.

Being a zero-waste volunteer is how I learned that recycling and compostable labels aren't regulated. Companies can put these messages on disposables even if they're not true. They do it to make us more comfortable buying their products.

- We feel better when we buy products if we think the packaging isn't bad for the environment.
- Laws and regulations that require packaging claims to be verified would help consumers make informed decisions, not misguided ones.

I want to make sure I'm doing good with my spare time and money, so I do a bit of research first. Though my preference is to volunteer and donate goods or food, I sometimes give financial support to organizations I know are effective.

No matter what works for you, first, investigate an organization before you support or promote it.

- Your Ecosia search engine can help you determine if an organization is helping the environment and how.
- Don't share posts on social media until you know they're good for the planet.

The things that affect change are often done at a local level. These focus on waste reduction, like eliminating single-use plastics or pushing for policies that hold corporate polluters accountable for emissions and cleaning up their messes.

- It can be easier to support local organizations and what they're accomplishing in your community.
    - You see the results in your local news and on your doorstep.

We have the power to require companies to take responsibility for their actions. We do this through our voices and our votes. We do it with local laws that say what corporations are allowed or not allowed to sell or build in our neighborhoods.

What to do and where to go when you have questions

You've probably heard the expression *knowledge is half the battle*. This has never been truer than it is with our fight against climate change. We need to stay informed and stop the misinformation that is bantered around the internet.

Tip 140 - Stay current on climate issues; find good resources to read and follow.

Look for podcasts where you can learn the facts and the science behind climate change. Learning the science is the best way to stay informed and know how to talk about the climate crisis.

- I found a blog post that listed 80 podcasts in 2022 on climate.
- Make sure what they talk about is backed by science.

Here are some of the organizations on the frontlines of the climate crisis. Many have free training for people who want to learn how to do more.

- The **All We Can Save Project**, allwecansave.earth, is helping people find ways to make a difference in our communities.
- **Climate One**, climateone.org, offers various podcasts and events to help engage us in today's climate issues and help us understand what needs to be done.
- The **Climate Reality Project**, climaterealityproject.org, provides training to help us fight for our planet and lead others to do the same.
  - A global organization started by Al Gore, they have online and in-person training available around the world.
- **Crowd Source Sustainability**, crowdsourcingsustainability.org, provides tools to learn how to stay positive and build a community around climate education.
  - Learn how to speak up at work and in your community.
- The **Earth Day organization** does more than put together events on Earth Day. They have year-round events and tools to help us learn sustainable living.
  - Visit EarthDay.org and sign up to receive their news and updates.
- **Earth911.com** has been my go-to for years to figure out how to recycle anything.
  - They're a great resource for more than recycling. I read their newsletters and listen to their podcast, *Sustainability in Your Ear*.
- **EcoWatch.com** publishes science-based informative articles on what's happening with our climate and planet.
  - Their newsletter is one of my favs.

- **Friends of the Earth**, friendsoftheearth.uk, is one of the oldest organizations fighting for people and the planet.
  - They're the UK's largest grassroots network.
- **Global Recycling Day**, globalrecyclingday.com, supporting efforts to improve recycling around the world.
- **Going Zero Waste** by Kathryn Kellogg, goingzerowaste.com, is packed with tips on how to reduce our waste habits.
  - She offers a free newsletter and a crash course on going zero waste. I always find Kathryn informative.
- **The Good Trade**, thegoodtrade.com, has a brief uplifting newsletter called The Daily Good with tips and resources to help you live sustainably.
- **Green Peace**, greenpeace.org, is a global network of organizations that use peaceful protest and creative communication to expose environmental problems.
  - Their website offers ways to act and learn about the latest issues in our fight to save the planet.
- **Grow Ensemble**, growensemble.com, promotes building a better world with helpful articles and interviews with experts.
- **Institute for Local Self-Reliance**, ILRS.org, helps you stay informed on what is available for you and your neighbors to build a more sustainable community.
- The NRDC.org newsletter, ***Top of Mind***, offers an in-depth understanding of new technologies that may or may not save our planet.
- **Ocean Conservancy**, oceanconservancy.org, provides evidence-based solutions for healthy oceans and wild places.
- **Sierra Club**, sierraclub.org, subscribe to their Green Life magazine to stay current on issues around our fight for the climate.
- **Treehugger.com**. They dig through the noise and get to the facts to help eco-friendly people make the right choices.
- **Union of Concerned Scientists**, ucsusa.org, is a US nonprofit founded over 50 years ago by scientists and students at the Massachusetts Institute of Technology (MIT).
  - They use scientific research and advocacy to find and support practical solutions.

- **WasteDive.com** has a newsletter I read because I'm interested in what's happening with waste management and recycling.
- Your local government website and newsletter to monitor sustainability action and initiatives in your area.
  - If your local government doesn't have this, ask them why.

Don't overdo it. You'll get overwhelmed. Pick two to start. Unsubscribe and add more as you find the ones you enjoy reading and are the most interesting to you.

- Put sustainability reading on your calendar, so you do it once a week.
  - Read either a book or a newsletter in your inbox, or both.
    - Your local library is a great way to try a book to see if you like it.

Set an Example - Teach Kids How Important the Planet Is

Unfortunately, the health of our planet has become political. Subsequently, parents need to be aware that not all schools are teaching kids about climate change. It's up to parents to make sure children know how to protect the planet.

Helping kids to understand what is happening to the planet and working with them to fight for it is the best way to handle everyone's anxiety over global warming.

Tip 141 - Encourage climate education and activism with your family and friends.

Young adults and teens are realizing the need for drastic change if they're going to have a livable planet. Encourage them to follow and join a youth-led climate activist group. Check the ones listed below and share them with your kids or search online to find a group you and your kids would like to join and support.

Don't force it on your teen or young adult family member; suggest and let them choose what they want to get involved in. No one puts forth their best effort when forced into something.

- **Action for the Climate Emergency (ACE)**, acespace.org, works to educate, inspire, and empower youth to get active about climate change.
- **Earth Guardians**, earthguardians.org, organizes and trains youth around the world to use civic engagement and the arts to help solve environmental issues.
- **Fridays for Future**, fridaysforfuture.org, is a youth lead strike held in March each year. Students and workers strike and march in the streets to increase awareness of the climate battle.
  - Imagine if everyone took PTO on the same day in March. Do you think companies might start to notice the need for change?
    - Maybe it could become a national holiday. I think the planet deserves one.
- **The Sunrise Movement**, sunrisemovement.org, is a youth lead organization fighting in the US for climate initiatives that will move us away from fossil fuels.
- **Zero Hour**, thisiszerohour.org, is a youth and women of color led movement that helps young people find their voice to speak up about climate and environmental justice.

Family time and climate education can go hand-in-hand.

Watch a documentary about the environment with the whole family. Schedule a regular time to do this once a month, or even once a week.

- Many of the organizations listed in **Tip 140** have resources for documentaries about what is happening to our climate.
  - Some reveal the truth about our fishing and livestock practices, such as *Seaspiracy* and *Eating Our Way to Extinction*. These are good to watch with older

children and young adults. They help us understand why we need to change our eating habits.

Kids are smart and appreciate it when you value their opinion. Ask them what organizations they like or think are doing good. They may know of some you haven't heard of.

- Maybe you'll find one you want to join together.
- Think of ocean clean-ups and groups to follow on social media.
  - Don't forget to verify posts before sharing them.
  - Let's keep the truth alive and strong.

Tip 142 - Support reduce, reuse, and recycle programs in schools.

A lot of school systems have recycling policies and programs, but they aren't very good at following them. Find out what kind of recycling is done in your schools and what they're doing to reduce waste.

- Do they print two-sided and use blank sides of paper for scrap?
- Do their cafeterias have reusable dishes and cookware?
- Does the school have a strategy to reduce the use of disposables?
- Do they encourage children to carry reusable lunch containers?
  - It's not hard to get kids to carry a lunch pail. That's how it was done until about 30 years ago when disposables became popular.
    - Build this habit with young children to help them be responsible for carrying things to school as they get older, like homework and textbooks.

There are many ways we can help schools do a better job at reducing, reusing, and recycling.

- Search for climate change education in your Ecosia search engine or on these sites:
  - The World Wildlife organization (WWF) has educational resources for schools at wwf.org and wwf.org.uk.
  - UNESCO is the United Nations Educational, Scientific and Cultural Organization, unesco.org, teaches "Peace, dignity, and equality on a healthy planet."
  - You'll find several universities have resources and educational materials available for people of all ages.
- The US EPA has this helpful guide, *Reduce, Reuse, Recycle Resources for Students and Educators*. Search for it at epa.gov and look for more resources under epa.gov/recycle.

Look for ways to make it fun!

There are so many fun family activities that are sustainable.

- Plant a regenerative garden. Let your kids take the lead to figure out how to do it. Depending on their age, you may have to help them find instructions online.
- Go back to Chapter 3 and look for crafts and reuse ideas you can do with your kids.

Tip 143 - Get books and games about climate change to help kids understand it.

There are writers creating fun stories for kids about the climate and why it's important. I've run across games, comics, and graphic novels that explain climate change in ways kids can understand. What a fun way to learn with your kids what needs to be done.

- *The Adventures of Captain Polo: The Climate Change Comic,* by Alan J. Hesse, is a comic book series about a polar bear sailing around the world learning about climate change.
- *The New Adventures of Captain Planet* is a comic book sci-fi series for older kids. There's also a tv show.

- NASA has a page dedicated to helping kids learn about climate. It includes games and easy information for kids to understand. Visit climatekids.nasa.gov.
- The site Recycling Center Near Me has a post with facts you can share with your kids at recyclingcenternear.me/recycling-facts-for-kids.
- The Climate Institute has a page dedicated to helping kids learn about our climate situation. Go to climate.org/climate-games.
- Games for Change is an organization working to engage kids with fun games and activities. Check out gamesforchange.org
- We Are Teachers has put together *15 Meaningful and Hands-On Climate Change Activities for Kids*. Find these at weareteachers.com/climate-change-activities.
- Check your local bookstore or online for climate-related stories for kids. There are a number of them available.
  - Every independent bookstore I've checked has them.

Whatever you do, don't live like an ostrich with your head in the sand. Learn the facts and help your kids learn them. Then, work together and get involved to help make the changes we need.

Activate the World – Time for a Big Push to Heal the Earth

Not everyone is cut out to be an activist. I surely didn't think I was one until our climate crisis changed that. Even though I didn't consider myself an activist for a large chunk of my life, I always volunteered and wanted to learn about my environment. So, in truth, I've always been an activist to some degree.

To Act = Activism

I love nature, but I've noticed how much it's changed since I was a kid, and not for the better. You don't need to be an activist marching in the streets to care for the environment.

There're many ways to help our planet heal beyond being responsible for our daily habits.

Pick something… anything related to climate you will enjoy doing. There's plenty to be done. Maybe it's cleaning up oceans and parks, fighting for our transition to renewable energy, or organizing zero-waste efforts at community events.

- Watch Dr. Ayana Elizabeth Johnson's Ted Talk from June 21, 2022, to help find your climate mojo.
- Whatever you find yourself interested in, work on it. Do it regularly. Set aside time once a week or at least once a month.

Beyond addressing your daily waste habits, find a cause and give some time or money to it on a regular basis. Then, keep adding more things you want to do and organizations to support.

- It's sad to say, but we need to be careful to who we give our time and money; we must do the research before we decide to support or follow an organization.
    - Some may be funded by the fossil fuel industry.
    - The good news is we have the internet and common sense to guide us.

Maybe you're really busy and only have a half hour this week. Use it to research organizations you want to support or read a sustainability newsletter and sign a few petitions.

- Do more during the weeks you aren't as busy.
- Keep trying to carve out more time for the planet.

Tip 144 - Getting outdoors is good for everyone's physical and mental health.

Most communities have events where you can volunteer to pick up trash in neighborhoods or along rivers. They usually have a newsletter listing opportunities to get involved. Get out in your community and look for one thing you can start doing now.

Unless you live in a very rural area, finding opportunities to help the environment isn't hard.

- Check your local government website.
    - o Are there neighborhood clean-ups going on?
    - o Search for tree planting events.
    - o Is there an e-waste event coming up that could use volunteers?
    - o Do they need zero-waste volunteers for the upcoming county fair?
        - Take initiative and find out how to make your community events zero waste.
        - Connect with others to build a network of people willing to learn how to support zero waste and volunteer at events.

Change happens because volunteers make it happen.

A common theme for most of us is *I'm too busy*. I get it. You probably picked up this book with some anxiety over whether you could come up with time to do these things.

Don't think of it as extra work you need to add to your to-do list. Instead, think of this as a shift in your attention.

- You set aside time to spend with your family and include some eco-friendly activities.
- You have to shop anyway, so why not develop green shopping habits?
- You're the one who decides if you're going to impulsively drive-thru a burger joint or head home for a meal.

Tip 145 - Mental health and meditation play a key role in healing our planet.

We often run from things that cause us anxiety or stress, and it's easy for climate change to be one of them. Climate anxiety can be very stressful because it can be hard to figure out what to do about it. The point of this book is to give you tools to help you feel more in control of that anxiety, not add to it.

Addressing how we escape from anxiety or stress-inducing topics is important. With the dawn of consumerism, retail therapy became an escape from the stresses of everyday life. It's a temporary fix for when we feel mentally or emotionally drained. The sad thing is, it's only a band-aid and doesn't repair what ails us.

Retail therapy really isn't therapeutic. Regrets over spontaneous buying can create more anxiety, especially when we recognize that it causes more waste.

- Find quiet practices to turn to when you're feeling stressed rather than running out to buy something you don't need and possibly can't afford.
    - Reading, exercise, or meditation can be great stress relievers.
- You don't need money to find mental health resources.
    - There are free podcasts and apps that help us deal with life's stresses.
        - Search your favorite streaming app for mental health, motivation, life skills, personal development, or a topic you want to address.
            - 10% Happier is a popular one, or
            - The Mel Robbins podcast is my favorite. I've learned many tools to help me fight anxiety, fear, and depression. All free.
    - Thank you, Mel.

Support the people who are helping us save the planet

There are several organizations and activists leading the way for change and a better planet. If your time is limited and you want to find ways to support these people, look for the ones you feel are making the biggest difference.

Visit the resources listed in this chapter and book. Then, when you find people or organizations you like, donate to their efforts so they can keep fighting for our planet.

Tip 146 - Follow, join, and support climate activists and programs making a difference.

It's easy to find climate activists that are making a difference. There are organizations and climate news outlets that follow and report on them and tell us how effective they are.

Find a few you like and follow them online. Then, donate to their efforts directly or to the organizations they work with. These resources might help you find your favorites.

- An article called *Meet 12 Climate Activists Changing The World* by Greenpeace.org gives a brief bio on each activist. It also tells what organizations they work with.
- Earthday.org has a list of *19 Youth Climate Activists You Should Be Following On Social Media*. Look up the article and read what they are involved in.
- Greta Thunberg. Keeping her activism alive is critical to our climate fight. She inspired the Fridays for Future youth climate movement. Follow her on Twitter at @GretaThunberg.
- And, my favorite, Dr. Ayana Elizabeth Johnson. I found her on the How to Save A Planet podcast and always look for ways to learn from her. Follow her on Twitter @ayanaeliza. She's ranked as the number one most influential climate change activist on social media.

Look for climate programs in your area that help you reduce waste and the use of disposables. If we don't support and buy goods and groceries through these programs, they can't stay in business.

- I once heard about a zero-waste grocer in the San Francisco Bay Area that delivered groceries in reusable packaging. Like milk delivery a hundred years ago, customers returned the containers to be washed and reused.
  - Sadly, when I moved to the Bay Area and went to add them as a resource for Chapter 3, I found out they had to close. Not enough support and customers.

It is critical we buy from and support the reuse industry.

If we don't support the efforts of businesses working to move us away from disposables, we'll end up stuck with the same-old-same-old. We can't afford that.

Where Do We Go From Here - The 100-foot View

Businesses, government regulations or deregulations, or our individual actions don't only cause global warming; all of these cause it, so these all need to change.

Tip 147 - Businesses big and small can do their part. These organizations can help.

If you own a business, take action to make it sustainable. Your customers will appreciate it, and it's good for business. People want to buy from companies with sustainable practices, and they're getting smart about finding them.

- Become a certified B-corp; visit bcorporation.net to learn how to develop a sustainable business model and become certified once you do.
- Global Recycling Network, grn.com, provides businesses with tools to reduce waste and helps companies share what they don't use rather than throwing it out.
- Plan A, plana.earth, uses data and science to help businesses decarbonize and develop sustainable practices.
- Give to environmental causes by donating at OnePercentForThePlanet.org.
- SkyWaterEarth.com will help your business to be sustainable and share that message with your customers and followers.
- Bluesign®, bluesign.com, helps apparel companies transition to sustainable materials and manufacturing practices. The bluesign® label means clothes meet the highest standards for sustainability.

Tip 148 - This transition will only happen if we all pitch in.

If we don't change our habits and make businesses and governments change, it is expected we will exceed a 1.5°C warming by 2030. This means top-down changes from big companies to you and me.

A few facts we need to stay grounded to:

1) We can have a healthy planet if we change our mindset and our habits **now**.
   a. This requires ditching our disposable lifestyles and waste habits.
2) When we speak out and work together, we can get governments and corporations working with us instead of against the planet.

Scientists are telling us we must do three things right away.

1. Fix agriculture
2. Stop overfishing
3. Ramp up clean energy

While these are big undertakings, these three actions could save our lives. We've learned ways to support these in this book. If we all start pushing for these changes and doing our part, we have a chance at reversing climate change.

Research shows that non-violent protests have been successful in creating large-scale change if at least 3.5% of the population participates.

- I like to think we are near that, but with Earth's population now at eight billion, we need at least 280 million activists to save the world.
  o It starts with you and me.
- Pay attention to waste habits.
  o Waste generation, energy generation... let's reduce it for future generations.
- Know what legislators are doing to protect the planet.

o   Stay informed and vote locally.
- Adopt a zero-waste mentality and set an example for your family and friends.
    o   Walk the talk and teach others how important reducing waste is to our future.

Look on the bright side - humans are capable of change

Progress has always meant moving forward. Taking what we've learned and making things better. It also means action.

"Change will not come if we wait for some other person or some other time. We are the ones we've been waiting for. We are the change that we seek."
Barack Obama

It doesn't take much to be an activist. Using tools like the Climate Action Now app to learn about current issues and send emails to your government representatives can be a strong form of activism. It's also a great way to become more involved and educated.

The goal is to make changes in our lives that help us feel good about how we tread on Earth. Oh, and if enough of us do it, we'll probably save humanity. No pressure :-)

1. People before profits. Corporations are the biggest cause of greenhouse gases.
    a. There are tools available to help any size business be more sustainable.
2. Taking time to volunteer in your community makes a difference.
    a. Get involved and commit to at least one environmental event each month.
3. Help children learn to be good stewards of the planet and learn with them.
    a. Make saving the planet a fun family activity.

4. Find ways to get involved outside of personal waste reduction. Get online and find informative resources.
   a. Follow activists making an impact and learn what needs to be done.
   b. Support effective movements, including youth activists.
5. Stay informed so you know what's happening and what you can do to help.
   a. The best ammunition against climate anxiety is action.
   b. We have hope for the future when our actions make it better.
   c. When we act, we feel in control.
6. Cleaning up the mess we made is critical right now.

We've got this!

About the Author

Jen Thilman has been a recycling and zero-waste volunteer for decades, which is how she learned that recycling doesn't always work. She has spent years working in green energy technology and studies sustainable living while she strives for it in her own life. Jen completed training from the Climate Reality Project in June of 2022.

Marrying her love for writing and the environment, Jen writes blogs about how to buy sustainable products, recycle everyday items, and save the planet. Her own blog on how to reduce, reuse, and rethink recycling can be found on her website at JenThilman.com.

Jen lives with her wirehaired terrier, Teddy Bear, in the San Francisco Bay Area. She writes non-fiction and fiction in many genres and formats, from short stories to poetry, as well as novels that include a post-climate apocalypse dystopian series.

Look for her on Instagram (jenthilman) and Twitter (@Jen_Thilman) to catch future posts of stories and book launch announcements.

HowExpert publishes how to guides on all topics from A to Z by everyday experts. Visit HowExpert.com to learn more.

About the Publisher

Byungjoon "BJ" Min is an author, publisher, entrepreneur, and the founder of HowExpert. He started off as a once broke convenience store clerk to eventually becoming a fulltime internet marketer and finding his niche in publishing. He is the founder and publisher of HowExpert where the mission is to discover, empower, and maximize everyday people's talents to ultimately make a positive impact in the world for all topics from A to Z. Visit BJMin.com and HowExpert.com to learn more. John 14:6

Recommended Resources

- HowExpert.com – How To Guides on All Topics from A to Z by Everyday Experts.
- HowExpert.com/free – Free HowExpert Email Newsletter.
- HowExpert.com/books – HowExpert Books
- HowExpert.com/courses – HowExpert Courses
- HowExpert.com/clothing – HowExpert Clothing
- HowExpert.com/membership – HowExpert Membership Site
- HowExpert.com/affiliates – HowExpert Affiliate Program
- HowExpert.com/jobs – HowExpert Jobs
- HowExpert.com/writers – Write About Your #1 Passion/Knowledge/Expertise & Become a HowExpert Author.
- HowExpert.com/resources – Additional HowExpert Recommended Resources
- YouTube.com/HowExpert – Subscribe to HowExpert YouTube.
- Instagram.com/HowExpert – Follow HowExpert on Instagram.
- Facebook.com/HowExpert – Follow HowExpert on Facebook.
- TikTok.com/@HowExpert – Follow HowExpert on TikTok.

Printed in Great Britain
by Amazon

61042702R00129